my revisi

WJEC GCSE

GEOGRAPHY

Includes full coverage of WJEC Eduqas GCSE (9–1) Geography A

Rachel Crutcher
Dirk Sykes

HODDER EDUCATION
AN HACHETTE UK COMPANY

The Publishers would like to thank the following for permission to reproduce copyright material.

Acknowledgements: p34 © Crown Copyright 2014 Office for National Statistics; **p36** Graph from Homes for London: The London Housing Strategy, 2013. Used with permission from Greater London Authority (GLA); **p38** Map data from Oxford City Council; **p78*b*** Based on data from GISS/NASA; **p122** United Nations Development Programme/Creative Commons Attribution 3.0 IGO; **p125*l,r*** United Nations, Department of Economic and Social Affairs, Population Division. World Population Prospects: The 2015 Revision.

Photo credits: p8*ml* SDM IMAGES/Alamy Stock Photo; **p8*tl*** Adam Burton/Alamy Stock Photo; **p8*bl*** Jeff Morgan 16/Alamy Stock Photo; **p8*tr*** travelib wales/Alamy Stock Photo; **p8*mr*** Ian Watt/Alamy Stock Photo; **p8*br*** Stephen Dorey Creative/Alamy Stock Photo; **p13** Paul Heinrich/Alamy Stock Photo; **p15** Dave Porter/Alamy Stock Photo; **p16*l*** Andy Owen; **p16*r*** B&JPhotos/Alamy Stock Photo; **p18** Joana Kruse/Alamy Stock Photo; **p20** cc-by-sa/2.0 - © Nigel Davies – geograph.org.uk/p/2685585; **p23** Ian Pilbeam/Alamy Stock Photo; **p24** Tomas Griger/Alamy Stock Photo; **p57** Robert Gilhooly/Alamy Stock Photo; **p60** keith morris news/Alamy Stock Photo; **p63** REUTERS/Alamy Stock Photo; **p64*a*** Joanne Moyes/Alamy Stock Photo; **p64*b*** Convery flowers/Alamy Stock Photo; **p64*c*** Leslie Garland Picture Library/Alamy Stock Photo; **p64*d*** geogphotos/Alamy Stock Photo; **p64*e*** Anglia Images/Alamy Stock Photo; **p66** Andy Owen; **p70** Natural Resources Wales. Contains public sector information licensed under the Open Government Licence v1.0; **p75** Robert Gray/Alamy Stock Photo; **p78*t*** David Hodges/Alamy Stock Photo; **85*t,b*** Met Office. Contains public sector information licensed under the Open Government Licence v1.0; **p92** Maciej Czekajewski/Alamy Stock Photo; **p102*t*** Ariadne Van Zandbergen/Alamy Stock Photo; **p102*b*** Noppasin Wongchum/Alamy Stock Photo; **p104** REUTERS/Alamy Stock Photo; **p111** Michael Dwyer/Alamy Stock Photo; **p121** Ordnance Survey (licence number 100036470); **p132*t*** AfriPics.com/Alamy Stock Photo; **p132*b*** Liam White/Alamy Stock Photo; **p136** Eye Ubiquitous/Alamy Stock Photo.

Orders: please contact Bookpoint Ltd, 130 Milton Park, Abingdon,
Oxon OX14 4SE. Telephone: +44 (0)1235 827720. Fax: +44 (0)1235 400454.
Email education@bookpoint.co.uk Lines are open from 9 a.m. to 5 p.m.,
Monday to Saturday, with a 24-hour message answering service. You can also
order through our website: www.hoddereducation.co.uk

ISBN: 978 1 4718 8740 6

First published in 2017 by
Hodder Education
An Hachette UK Company
Carmelite House, 50 Victoria Embankment
London EC4Y 0DZ

www.hoddereducation.co.uk

Impression number	10 9 8 7 6 5
Year	2021 2020 2019

Cover photo © Aurora Photos/Alamy
Illustrations by Gray Publishing
Produced and typeset in Bembo by Gray Publishing, Tunbridge Wells, Kent
Printed in Spain

A catalogue record for this title is available from the British Library.

Get the most from this book

This revision guide has been written to accompany the WJEC GCSE (A*–G) Geography and WJEC Eduqas GCSE (9–1) Geography A specifications to help you get the best possible result in your examinations.

This book aims to give you the essentials that should serve as a reminder of what you will have covered in your course and allow you to bring together your own learning and understanding.

Everyone has to decide his or her own revision strategy, but it is essential to review your work, learn it and test your understanding. These revision notes will help you to do that in a planned way, topic by topic. Use this book as the cornerstone of your revision and don't hesitate to write in it – personalise your notes and check your progress by ticking off each section as you revise.

Tick to track your progress 

Use the revision planner on pages 4–7 to plan your revision, topic by topic. Tick each box when you have:

- revised and understood a topic
- tested yourself
- practised the exam questions and gone online to check your answers.

You can also keep track of your revision by ticking off each topic heading in the book. You may find it helpful to add your own notes as you work through each topic.

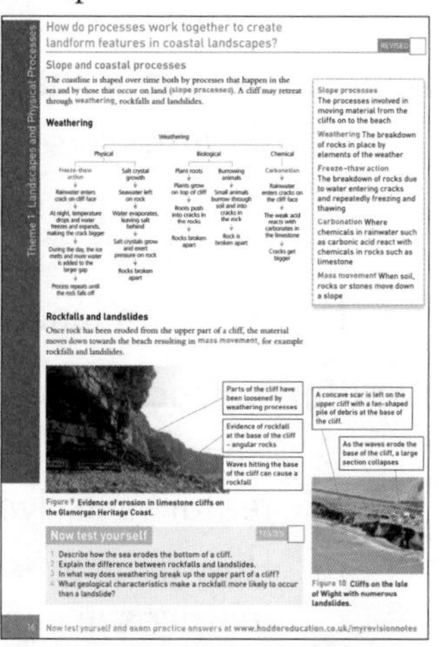

Theme 1: Landscapes and Physical Processes

How do processes work together to create landform features in coastal landscapes? REVISED

Slope and coastal processes

The coastline is shaped over time both by processes that happen in the sea and by those that occur on land (slope processes). A cliff may retreat through weathering, rockfalls and landslides.

Weathering

Slope processes The processes involved in moving material from the cliffs on to the beach

Weathering The breakdown of rocks in place by elements of the weather

Freeze–thaw action The breakdown of rocks due to water entering cracks and repeatedly freezing and thawing

Carbonation Where chemicals in rainwater such as carbonic acid react with chemicals in rocks such as limestone

Mass movement When soil, rocks or stones move down a slope

Rockfalls and landslides

Once rock has been eroded from the upper part of a cliff, the material moves down towards the beach resulting in mass movement, for example rockfalls and landslides.

Figure 9 Evidence of erosion in limestone cliffs on the Glamorgan Heritage Coast.

Figure 10 Cliffs on the Isle of Wight with numerous landslides.

Now test yourself

1. Describe how the sea erodes the bottom of a cliff.
2. Explain the difference between rockfalls and landslides.
3. In what way does weathering break up the upper part of a cliff?
4. What geological characteristics make a rockfall more likely to occur than a landslide?

16 Now test yourself and exam practice answers at www.hoddereducation.co.uk/myrevisionnotes

Features to help you succeed

This guide contains features intended to help you to *actively* work through your revision schedule.

Revision activities

These activities have been designed to focus your revision. If you can complete these activities, congratulate yourself and give yourself a reward. If you find some of the activities challenging, read over your notes again and speak to your teacher if you need further help.

Examples

It is vital that you can give specific examples, particularly in the questions requiring extended answers. This demonstrates the detailed knowledge examiners are looking for to award the higher grades. Often, your example does not require lots of detail but should be used to back up a point being made rather than simply recounted out of context.

Now test yourself

These activities are designed to help you decide if you fully understand a topic and are exam ready. If you struggle, then read over your notes again and perhaps ask to see your teacher or bring the topic up in revision classes.

Exam practice

These questions closely resemble the style of the question that you will face in the examination. Use them to consolidate your revision and practise your exam skills.

Exam tips

Expert tips are given throughout the book, including identifying common mistakes and suggesting strategies for getting the best out of the time you have in the examination room. These will therefore help to boost your final grade.

Definitions and key terms

Here you will find some of the specialist geography terminology that you need to know. Key terms are highlighted in bold throughout the book. Clear, concise definitions are provided where the essential key terms first appear.

Online

Go online to check your answers to the 'now test yourself' questions and the 'exam practice' questions at **www.hoddereducation.co.uk/myrevisionnotes**

My revision planner

Changing Physical and Human Landscapes

Theme 1 Landscapes and Physical Processes

Theme 2 Rural–Urban Links

Theme 3 Tectonic Landscapes and Hazards

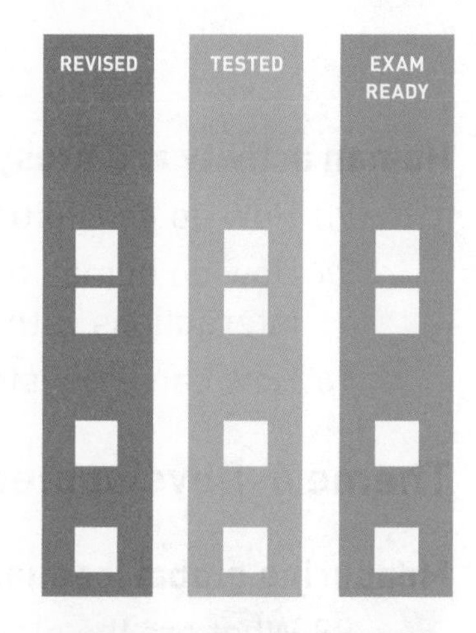

Theme 4 Coastal Hazards and their Management

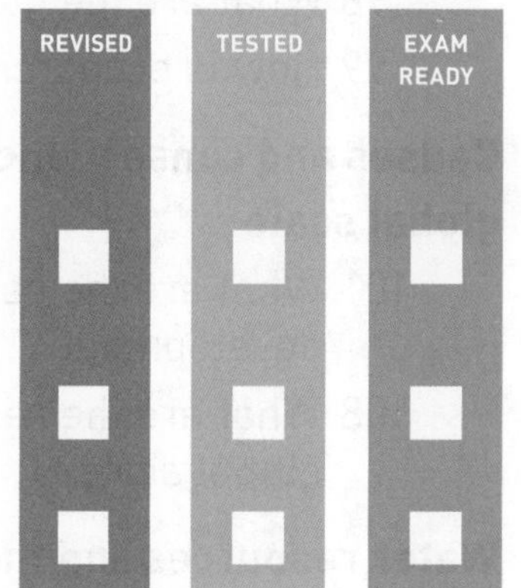

Environmental and Development Issues

Theme 5 Weather, Climate and Ecosystems

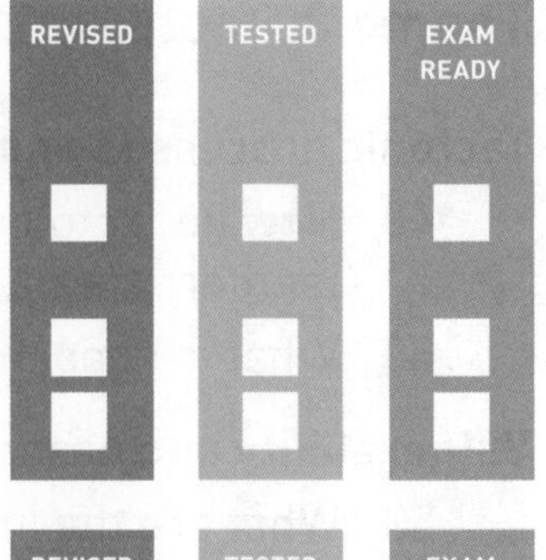

Theme 6 Development and Resource Issues

Theme 7 Social Development Issues

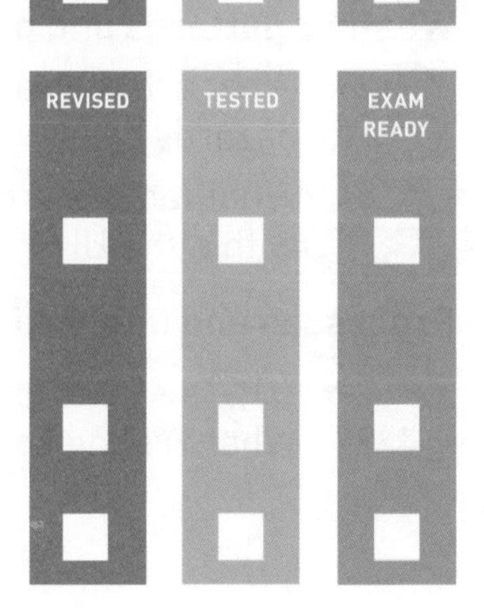

Theme 8 Environmental Challenges

Now test yourself answers and exam practice answers at www.hoddereducation.co.uk/myrevisionnotes

Getting to know the specifications at www.hoddereducation.co.uk/myrevisionnotes

Fieldwork enquiry at www.hoddereducation.co.uk/myrevisionnotes

Theme 1 Landscapes and Physical Processes

Distinctive landscapes

What makes landscapes distinctive?

REVISED

Landscapes are made up of different features and landforms. How these features and landforms combine is what gives a landscape its special or **distinctive** appearance. You need to be able to identify and locate examples of distinctive landscapes of the UK, including:

- **upland** and **lowland** areas
- river and coastal landscapes.

Upland A landscape that is hilly or mountainous

Lowland An area of land that is lower than the land around it

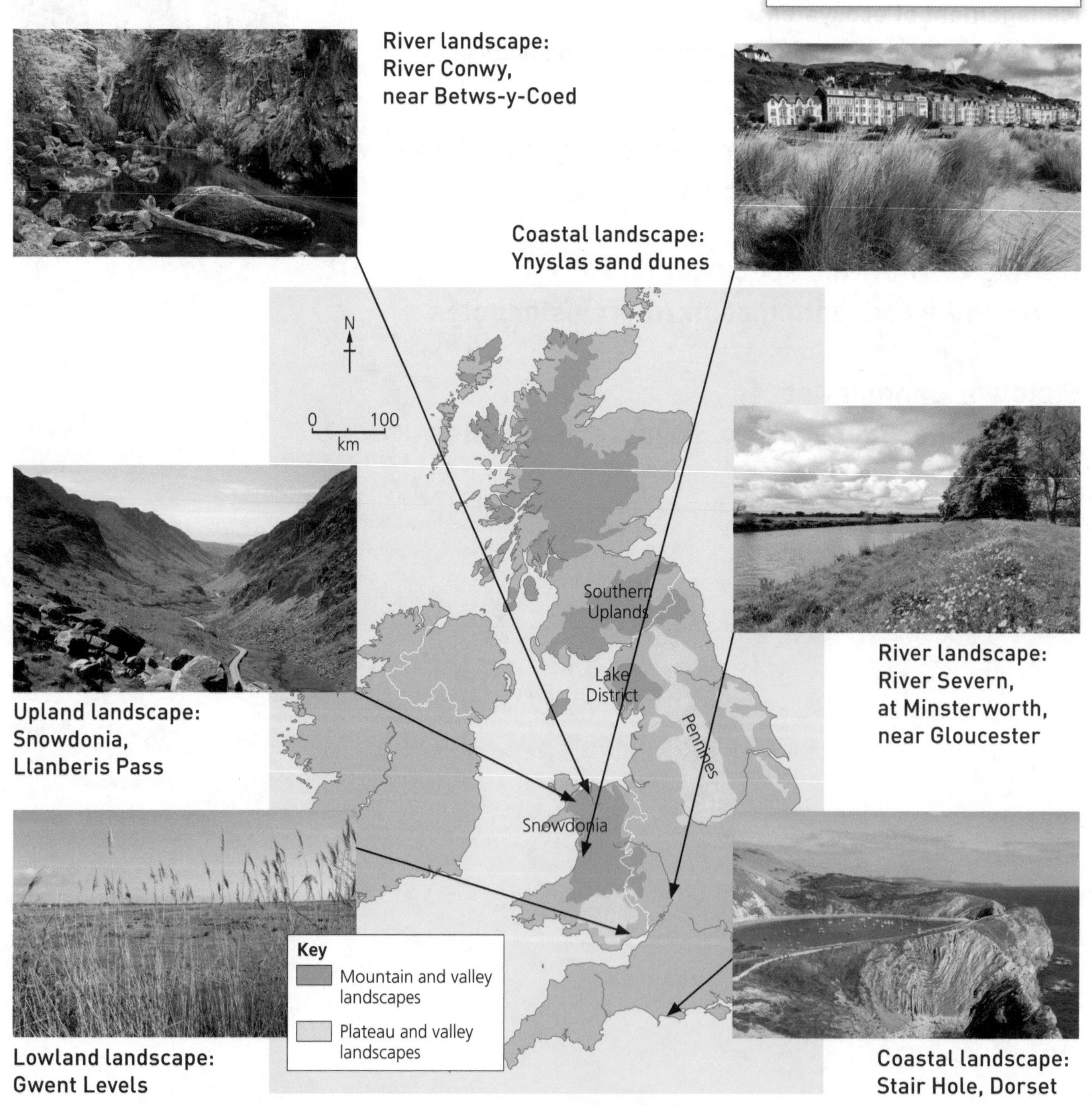

Figure 1 Main types of landscapes in the UK.

Example of a distinctive landscape: Snowdonia

- **Geology:**
  - Diverse upland landscape shaped by volcanic eruptions and extensive glaciation.
  - Numerous glacial features such as corries, U-shaped valleys and arêtes.
  - Mountain range that drops quickly to the sea.
  - Contains the highest mountain in Wales – Snowdon, 1085 m.
- **Land use:**
  - The landscape in many areas has been shaped by slate mining.
  - Large areas of agriculture (mainly pastoral) and forestry.
  - National Park status – attracts thousands of tourists each year, which has led to the growth of B&Bs, camp sites and other tourist facilities.
- **Vegetation:**
  - Diverse range of both plant and animal life due to the varied physical environments and habitats found.
  - Unique and rare species such as the Snowdon lily and the Snowdon beetle.
  - Large areas of natural mixed deciduous forests including species such as Welsh oak and birch; and planted coniferous trees which are often harvested.
- **People and culture:**
  - Snowdonia has a rich cultural history with many World Heritage sites including Celtic shrines and fortresses.
  - Welsh language widely spoken.

Exam tip

When asked to consider factors that make landscapes distinctive, remember to discuss physical and human features such as:

- geology
- people and culture
- vegetation
- land use.

Revision activity

Create a spider diagram to illustrate the distinctive characteristics of a landscape in the UK that you have studied. Remember to include both physical and human features.

Now test yourself

TESTED

1. What do you understand by the term 'distinctive landscape'?
2. Name four types of distinctive landscape.
3. List four factors that influence landscapes.
4. Explain why Wales and the UK have such a range of distinctive landscapes.

How are physical landscapes affected by human activity?

REVISED

Human activity can have both positive and negative impacts on the natural environment, for example:

- **Positive:** visitors to the countryside bring benefits to rural economies through the money they spend.
- **Negative:** **visitor pressure** may adversely affect the landscape and local communities.

For **honeypot sites**, where **carrying capacity** is likely to be regularly exceeded, this poses **environmental challenges**.

Example of environmental challenges in Snowdonia National Park

Snowdonia is a glaciated upland landscape, with a population of about 25,000. Nearly 4.3 million people visit the **National Park** each year, spending £396 million.

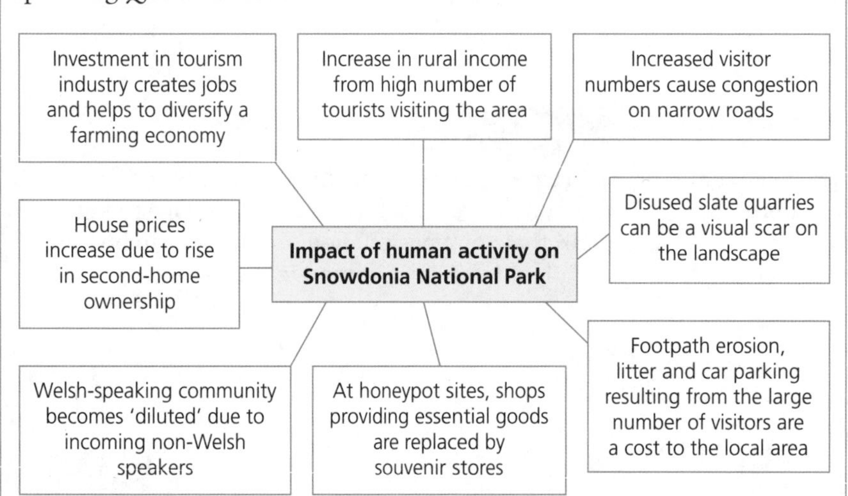

Visitor pressure The increased impact on the landscape, resources and services of an increased number of people due to tourism

Honeypot site A place of special interest that attracts tourists

Carrying capacity The maximum population size that an environment can sustain

Environmental challenges Problems caused by human use of the natural landscape or resources

National Park An area which is protected because of its beautiful countryside, wildlife and cultural heritage

Now test yourself

TESTED

1. Look at the example shown of impacts from human activity in the Snowdonia National Park. Identify which impacts are positive and which are negative.
2. Explain why honeypot sites may help to ease visitor pressure in surrounding parts of the distinctive landscapes.

Exam practice

1. Outline the human and physical features of a distinctive landscape that you have studied. [4]
2. What are honeypot sites and how can they help to protect distinctive landscapes? [4]
3. For a distinctive landscape that you have studied, discuss whether the impacts of human activity are largely positive or negative. [8]

Revision activity

For a distinctive landscape that you have studied, draw a spider diagram to show the impacts of visitor pressure. Use colour to code these impacts into positive and negative.

Exam tip

Make sure you know what makes different landscapes distinctive: upland, lowland, river and coastal. You also need to be able to describe:

- the type of landscape it is
- its location (where it is)
- its smaller-scale human and physical features.

How can landscapes be managed?

REVISED

Many distinctive landscapes in the UK are designated **Areas of Outstanding Natural Beauty (AONB)** or National Parks. These areas are visited by large numbers of people who, through sheer numbers, may cause damage to the natural landscape. Visitors must be managed in a way that minimises their impact on the landscape, and any damage caused must be repaired.

Area of Outstanding National Beauty (AONB) A part of the countryside that is designated for conservation due to its natural beauty

Examples of strategies to manage landscapes: the Gower and Brecon Beacons

Management of visitors in the Gower AONB

- Designated footpaths provide access for visitors but also protect sensitive areas.
- Detailed information boards at popular visitor sites inform visitors of the unique features of the location, for example Oxwich Bay Nature Reserve.
- Clearly marked car parks reduce parking on grass verges, which can damage hedgerows and cause congestion on narrow roads.
- Strict control over planning and building within the area ensures that new developments are restricted, and that extensions or changes to land use do not spoil the natural beauty of an area.

Footpath maintenance in the Brecon Beacons National Park

- Volunteer recruitment, for example local environmental groups or ecotourists help National Park wardens to repair footpaths and walls that have been damaged.
- Logistical operations, for example a helicopter is used to carry footpath materials due to the remote location and weight of the materials used.
- Footpaths are replaced with hard-wearing materials such as stone.
- Once the footpath has been rebuilt, vegetation can be restored on either side of the path to maintain the unique flora of the area.

Now test yourself

TESTED

1. List the impacts that visitors have on the natural landscape.
2. Give one method by which these impacts can be addressed.
3. In what way can increased tourism benefit a natural landscape?
4. Create a table to show the advantages and disadvantages of the ways in which visitors can be managed in a distinctive landscape that you have studied.

Exam practice

For a distinctive landscape that you have studied:

1. Identify how that landscape has become damaged due to visitors. [2]
2. Explain why this damage has occurred due to increased visitor numbers. [4]
3. Describe ways in which the landscape has been repaired. [4]

Exam tip

When describing ways in which visitor numbers can be managed, remember to link the management technique to how it minimises the impact that visitors have on the landscape.

Landform process and change

How do processes work together to create landform features in river landscapes?

REVISED

Fluvial processes

River landforms change over time due to **fluvial erosion**, **transportation** and **deposition**.

Fluvial Referring to a river and its landforms

Erosion The wearing away of the land

Transportation The movement of material by the flow of water

Deposition The dropping of the material carried by the river

Bed load The material carried by the river being bounced or rolled along its bed

Meander A bend in the river formed by lateral (sideways) erosion

Fluvial erosion

The type of erosional process that occurs in a river will depend on a number of factors including the velocity of the water and the rock and soil type of the channel. The erosional processes of the river channel include:

- **Abrasion**: stones and material carried by the river hitting the river bed and banks, wearing them away.
- **Hydraulic action**: the sheer force of water hitting the river bed and banks, compressing air in gaps in the soil and rock which causes material to be washed away.
- **Solution**: the slightly acidic river water dissolves chalk and limestone rocks which are made from calcium carbonate.

The erosional processes of the river **bed load** include:

- **Attrition**: stones carried by the river collide together and are broken down, becoming rounder and smaller.
- **Abrasion**: stones and material carried by the river hitting the river bed and banks become eroded themselves to become rounder and smaller.

Transportation

The river transports (moves) its load in a number of ways which depend on the speed of flow and the weight of the load.

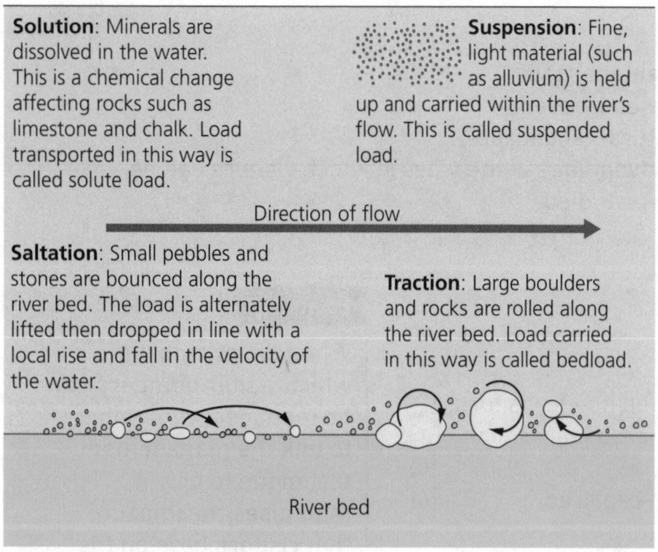

Figure 2 River transport processes.

Deposition

A river deposits material when the speed of flow is too slow for it to carry the load. This may happen:

- Where there has been a lack of rainfall, so there is less water moving in the river channel.
- On the inside of a **meander** because the majority of the water is on the outside of the bend. Therefore, the water on the inside of the bend is moving slowly and cannot transport load.
- At the mouth of the river, where the river water flows against the direction of the sea.

You also need to know the definitions of the following terms: **abrasion, attrition, hydraulic action, solution**

Now test yourself

TESTED

1 Describe the processes by which a river erodes its channel.
2 What are the factors that determine which method of transportation material is moved by? Give examples of the conditions when each type of transportation may occur.

How river landforms develop

V-shaped valleys, **waterfalls**, **gorges**, **floodplains** and **meanders** are all river landforms shaped by fluvial processes.

V-shaped valleys

V-shaped valleys are found in the upper course of a river valley, where the river is usually small and the land is steep.

> **Exam tip**
>
> When asked to explain the formation of a river landform, remember to describe the sequence of how the landform develops as well as explaining the processes involved.

Example of a V-shaped valley: Brecon Beacons

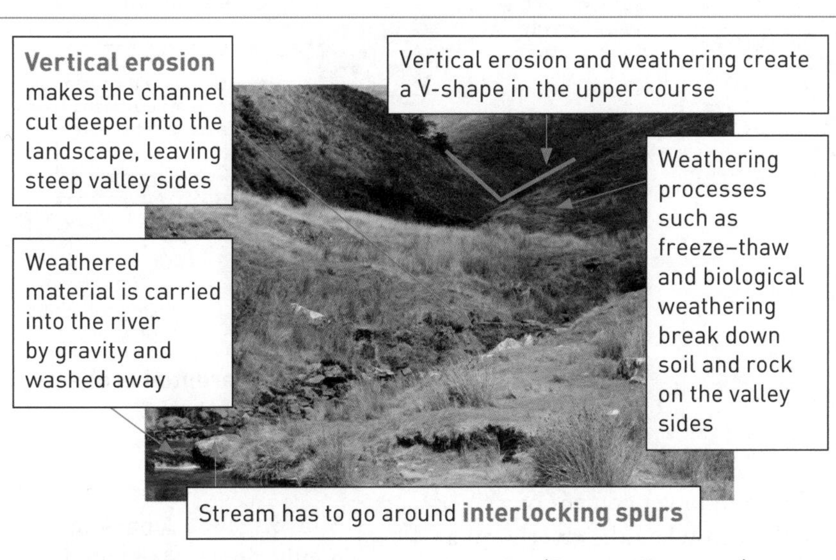

Figure 3 Blaen Taff Fawr Stream on Corn Du (Brecon Beacons).

V-shaped valley A narrow valley with steep sloping sides found in the river's upper course

Waterfall Water falling from a higher level to a lower level due to a change in rock structure or as a result of glacial erosion

Gorge A steep-sided narrow valley formed by a retreating waterfall

Floodplain A flat piece of land on either side of a river forming the valley floor

Meander A bend in the river formed by lateral (sideways) erosion

Vertical erosion Erosion of the river channel that results in its deepening rather than widening

Interlocking spurs Hard, resistant rocks that a river cannot easily erode and therefore the river goes around them

Plunge pool A deepened part of the river bed at the base of the waterfall caused by the impact of the falling water

Waterfalls and plunge pools

Waterfalls can be formed in one of two ways: by **glacial erosion** or by **differential erosion**.

Glacial erosion: where waterfalls have formed due to the erosive power of a glacier during the ice age. Glaciers carved steep valleys into the landscape, often hanging above one another. Once the glacier melted, water drains from the smaller valleys and falls into the larger ones.

Differential erosion: where waterfalls are formed due to a change in rock structure (hard and soft rock), which leads to the river bed being eroded at different rates:

- As the river bed crosses on to soft rock from hard rock it is eroded (hydraulic action and abrasion) at a faster rate and a step is created.
- As the water 'falls', hydraulic action continues to erode the rock underneath the hard rock as it splashes against it.
- As the soft rock is further eroded, the overhang becomes too heavy and the rock collapses, causing the position of the waterfall to retreat upstream.
- A **plunge pool** is created underneath the waterfall due to the sheer force of the water hitting the river bed and the abrasion caused by the rocks from the overhang being moved by the water.

> **Exam tip**
>
> You may be asked about a landform's associated smaller-scale features, so make sure you know how the plunge pool is formed.

Gorge

A gorge is a steep-sided narrow valley with a river running along the bottom of it. A gorge is formed when a waterfall collapses and retreats upstream, and has characteristic vertical sides.

> **Exam tip**
>
> Remember a **slip-off slope** (point bar) is a smaller-scale feature of a meander and you may be asked to explain its formation.

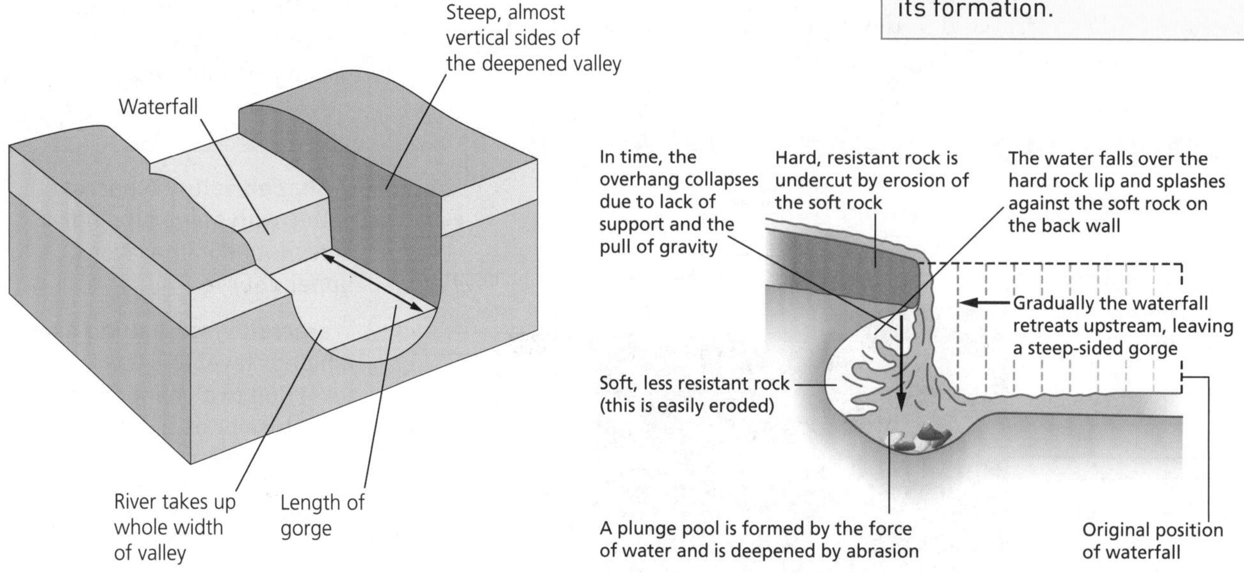

Figure 4 Block diagram of a gorge.

Figure 5 Waterfalls formed by differential erosion.

Meanders

Meanders are usually found in the middle and lower courses of a river valley. They are bends in the river usually seen when the river is on a wide floodplain. Caused by both erosion on the outside of the bank and deposition on the inside of the bank, meanders can often be seen to 'move' or 'migrate' across the valley floor as the river channel changes position.

> **Slip-off slope** A bank of gently sloping deposited material found on the inside bend of a meander

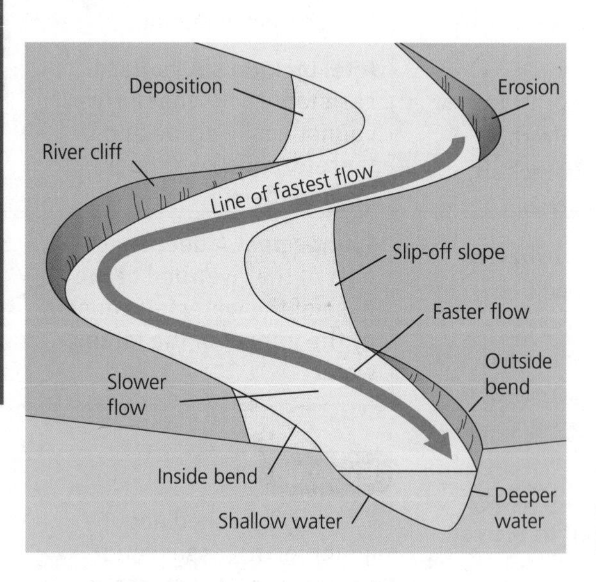

Figure 6 Characteristics of a meander.

> **Revision activity**
>
> Make two flashcards, one for the outside bank of a meander and one for the inside bank. On these cards write bullet points of the key features of both the water and the bank at these points and remember to use all this information in an explanation of the landform formation.

Now test yourself

TESTED ☐

1. Describe how a V-shaped valley is formed.
2. How do river processes lead to the formation of a waterfall?
3. Explain how the smaller-scale feature of a plunge pool is formed.

Exam practice

Explain two different ways that waterfalls can form in Wales. [6]

Floodplain

- When the river floods, the floodplain becomes covered with water.
- As the water is shallower on the land than it is in the river, material (silt) is deposited.
- The silt makes the soil fertile.

Figure 7 A meander on the River Severn.

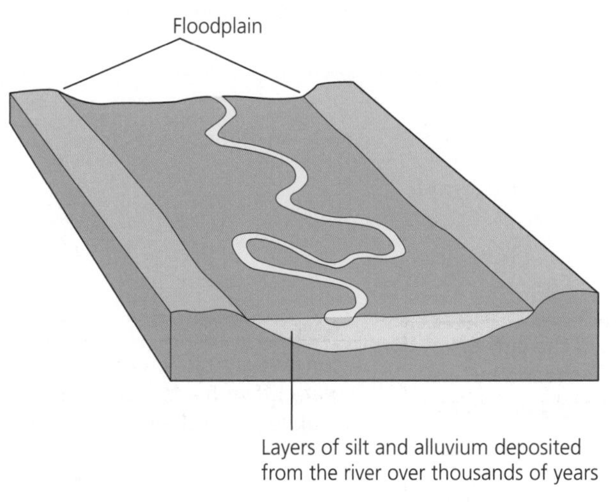

Figure 8 A floodplain.

Revision activity

Make an individual flashcard for each river landform that you have studied. For each card you should:

- include a diagram of the landform
- label the diagram with key features
- bullet point the processes involved in its formation
- give a located example that you have studied.

Now test yourself

TESTED

1. Annotate a diagram to explain how a floodplain is formed.
2. Why do meanders migrate across floodplains?

Exam practice

1. Give two factors that influence which method of transportation a river's bed load is moved by. [2]
2. Describe the formation of a V-shaped valley. Use a diagram to help your answer. [4]
3. 'Erosional processes are the most important factor in the formation of river landforms.' To what extent to do you agree with this statement? [8]

How do processes work together to create landform features in coastal landscapes?

REVISED

Slope and coastal processes

The coastline is shaped over time both by processes that happen in the sea and by those that occur on land (**slope processes**). A cliff may retreat through **weathering**, rockfalls and landslides.

Slope processes The processes involved in moving material from the cliffs on to the beach

Weathering The breakdown of rocks in place by elements of the weather

Freeze–thaw action The breakdown of rocks due to water entering cracks and repeatedly freezing and thawing

Carbonation Where chemicals in rainwater such as carbonic acid react with chemicals in rocks such as limestone

Mass movement When soil, rocks or stones move down a slope

Weathering

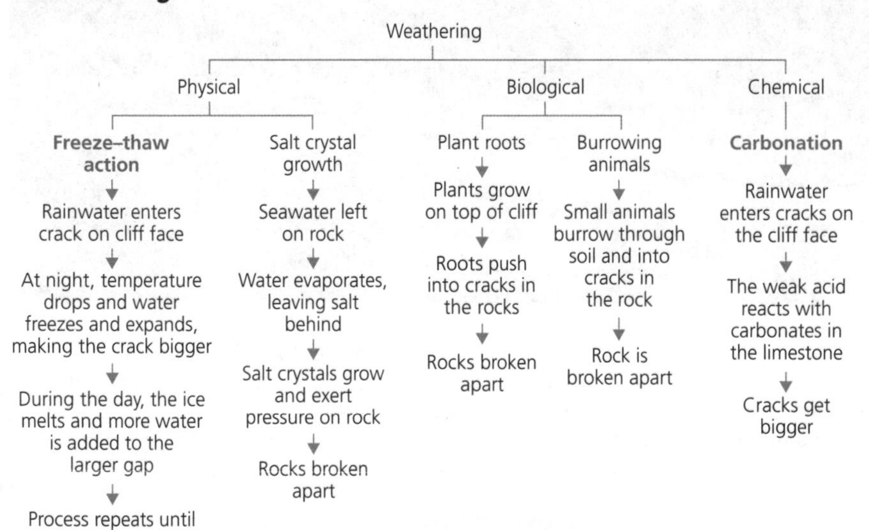

Rockfalls and landslides

Once rock has been eroded from the upper part of a cliff, the material moves down towards the beach resulting in **mass movement**, for example rockfalls and landslides.

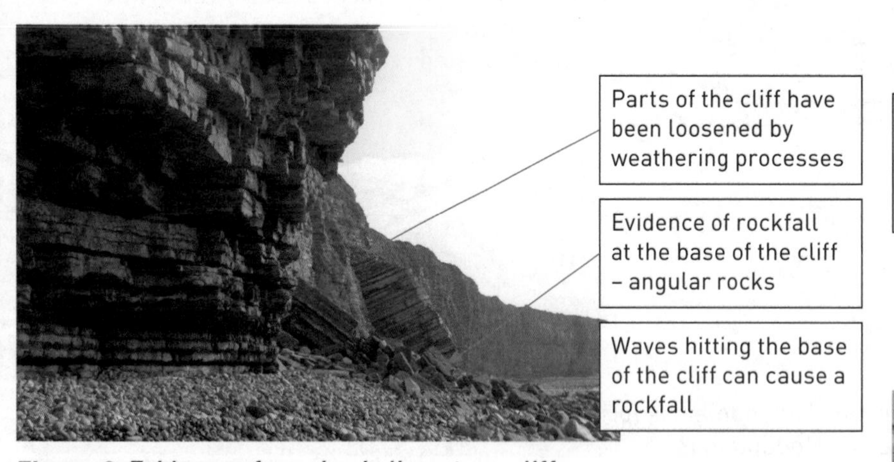

Figure 9 Evidence of erosion in limestone cliffs on the Glamorgan Heritage Coast.

Figure 10 Cliffs on the Isle of Wight with numerous landslides.

Now test yourself

TESTED

1 Describe how the sea erodes the bottom of a cliff.
2 Explain the difference between rockfalls and landslides.
3 In what way does weathering break up the upper part of a cliff?
4 What geological characteristics make a rockfall more likely to occur than a landslide?

Coastal erosion

The processes that erode the cliff are:

- **Hydraulic action**: the force of waves crashing into cliffs. Air trapped in the cracks is compressed, which breaks up the rock.
- **Abrasion**: waves hurl sand and pebbles against the cliff, which wears the land away.
- **Solution**: salt water dissolves rocks made of calcium carbonate.

The processes that erode the beach material are:

- **Abrasion**: waves hurl sand and pebbles against the cliff, which wears the land away.
- **Attrition**: pebbles are rolled back and forth. They collide with each other which makes them smaller and rounder, eventually turning them into sand.

Coastal transportation and deposition

Once the eroded material, called **sediment**, falls into the sea, it will be transported by the power of the waves and currents along the coastline by **longshore drift**.

Sediment The material carried by the sea

Longshore drift The process by which sediment is moved along the coastline

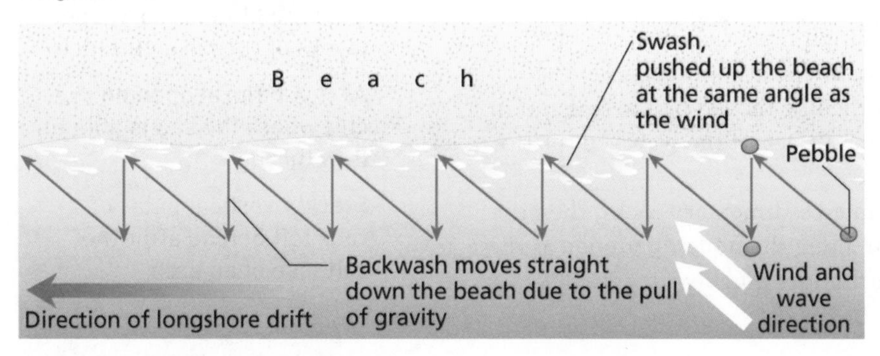

Figure 11 Longshore drift.

Example of sand dunes created by longshore drift in Ynyslas

Ynyslas sand dunes are located on the west coast of Wales in Ceredigion and provide an excellent example of how a sand dune system can be created by longshore drift. The map explains why the sand dunes are found here, as well as the development of an offshore bar.

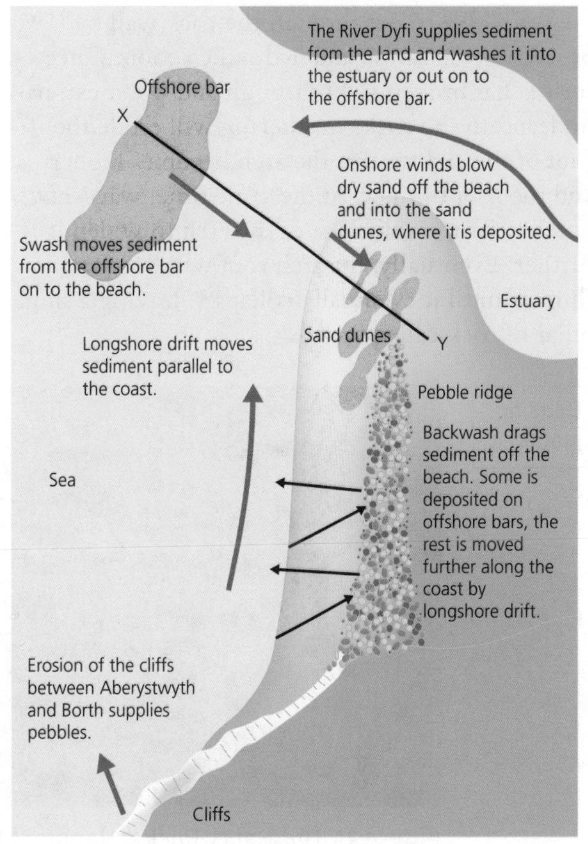

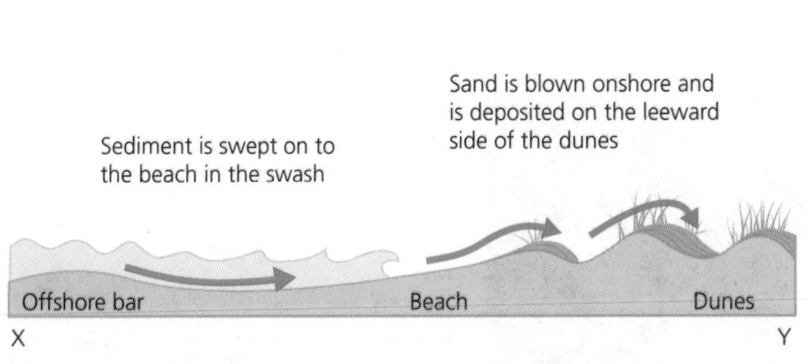

Figure 12 The transportation of beach sediment at Borth and Ynyslas on the Ceredigion coast.

Revision activity

Draw a mind map to show all the different ways in which a coastline can be eroded.

Distinctive coastal landforms

The coastal environment has both large-scale and smaller-scale features.

Headlands and bays

- A **headland** is an area of land that juts into the sea and is formed due to harder, more resistant rock being eroded more slowly.
- A **bay** is formed between the headlands due to softer, less resistant rock which erodes more quickly. Beaches often form in sheltered bays.

Cliffs and wave-cut platforms

A **wave-cut platform** is formed when a cliff face is eroded by the sea:

- As the waves pound the base of the cliff, hydraulic action and abrasion cut a **wave-cut notch** into the base of the cliff which makes the cliff vulnerable to collapse.
- With continued erosion at every high tide, the wave-cut notch will eventually make the cliff unstable and lead to its collapse.
- The material from the cliff will then be moved by the sea, and in doing so abrasion will smooth the surface of the wave-cut platform left behind.
- If the cliff is made from well-jointed sedimentary rocks, then the wave-cut notch will often occur along the **bedding planes** as these are a weak point and will erode much more quickly.

Headland An area of land that juts into the sea

Bay A recessed area of coastline often found between two headlands

Wave-cut platform A coastal landform made of a rocky shelf in front of a cliff

Wave-cut notch A slot with overhanging rocks that has been cut into the bottom of a cliff by wave action

Bedding plane Clearly seen layers of rock in a cliff face

Arch A natural opening in a cliff where the sea is able to flow through

Stack A vertical pillar of rock left behind after the collapse of an arch

Arches and stacks

Arches and **stacks** form in headlands made from rock which is relatively resistant to erosion, for example limestone. Arches form when two caves are created on either side of a headland. Over time, the sea is able to cut through the back wall by processes of abrasion and hydraulic action. Once the sea has broken right through and water can pass underneath the rock, weathering will erode the roof of the arch so that the arch becomes higher and the roof thinner. At the same time, wave-cut notches form on the base of the arch to widen it further. Eventually, the arch roof will become thinner until it eventually collapses, leaving a single pillar of rock called a stack.

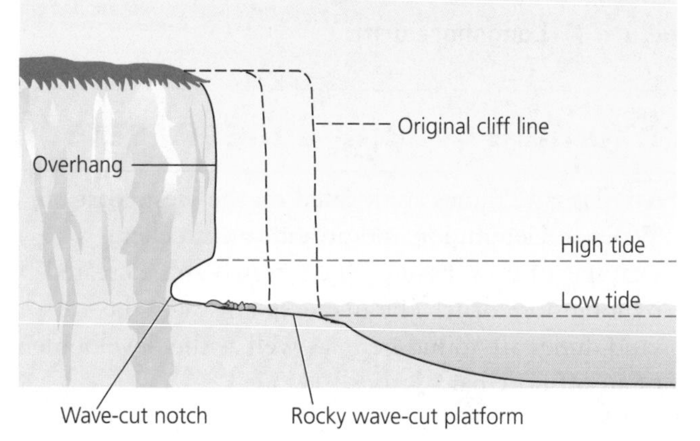

Figure 13 The formation of cliffs and wave-cut platforms.

Figure 14 Old Harry Rocks, Handfast Point, near Swanage, Dorset.

Revision activity

Create a flashcard for each large-scale coastal landform. Use a diagram to explain its formation and make note of any smaller-scale features that may be present.

Beaches and spits

Beaches and spits are formed when the **swash** is stronger than the **backwash** and deposition occurs:

- A **beach** is a build-up of sand, shingle and pebbles deposited by waves.
- Longshore drift transports beach material along the coast. Where the coast changes direction, for example at a river mouth, beach material is carried out to sea. This creates a new strip of land which projects out into the sea and remains attached to the land at one end, called a **spit**. Spurn Point on the Holderness coastline at the mouth of the River Humber is an example of a spit.
- Fine silts and sands that are transported by the river are deposited at the river mouth and form an **offshore bar**. The mouth of the Dyfi Estuary (see the longshore drift example) is an offshore bar. This material can then be washed onshore by the swash action.

Swash The movement of water up the beach as a wave breaks

Backswash The flow of water back into the sea after a wave has broken on to the shore

Beach Created by deposition (usually sand, shingle or pebbles) and lies between the high water mark and the low water mark

Spit A sand or shingle beach that is joined to the land but projects outwards into the sea in the direction of the prevailing wind

Offshore bar An area of deposition that is slightly off the coastline in the estuary of a river

Example of Spurn Point, Holderness coastline

- Spurn Point is a narrow spit on the east coast of the UK which has formed across part of the Humber Estuary.
- The spit is 4.8 km long and as narrow as 46 m in places.
- Sand and shingle are moved along the Holderness coastline by longshore drift to the mouth of the Humber River.
- Deposition occurs in the more sheltered water and a spit develops.
- As more sand is deposited, colonising plants such as marram grass begin to grow, which stabilises the spit further.
- Longshore drift continues along the spit, which increases its length.

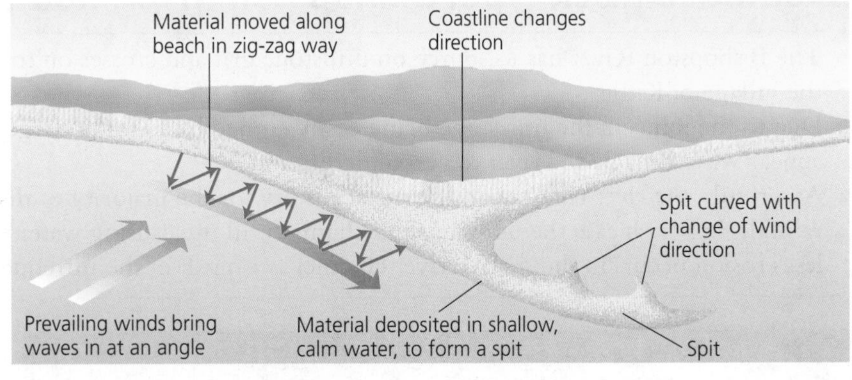

Figure 15 **Formation of a spit.**

Revision activity

Make a revision card for a located coastal environment that you have studied, and include on it:

- a map of the location
- what coastal landforms are found there
- why those landforms are found there
- the direction of longshore drift (if appropriate)
- the source of sediment supply (if appropriate).

Rock pools

- **Rocks pools** are small hollows in rocks found at the coastline such as in a wave-cut platform.
- At high tide the pools are covered by the sea, and at low tide some seawater remains in the hollow, creating a rock pool.
- The rock pools are enlarged by the process of abrasion at high tide as small rocks within the pool whirl around due to the movement of the waves and gradually increase the size of the hollow.

Rock pool A pool of seawater between shoreline rocks

Now test yourself

TESTED

1. What is the difference between a spit and an offshore bar?
2. Explain where the material comes from to create spits and offshore bars.
3. Why do rock pools develop?
4. Explain why some spits may become eroded over time.

Exam practice

For a located coastal environment in the UK, explain the processes that have created the landforms specific to that environment. [6]

What factors affect the rates of landform change in river and coastal landscapes?

REVISED

Geology, climate and human activity will affect the rate of landform change (how fast the change happens) in both river and coastal environments.

Geology

The type of rock that is being eroded and the way in which the rock types are laid down will affect the rate of change.

Example of rock type: Bishopston River, South Gower

- The Bishopston River has its source on millstone grit and crosses on to carboniferous limestone near the village of Kittle.
- Due to the joints in the limestone being easily eroded by solution, large **caverns** and **sink holes** appear which enable the river to run underground.
- As a result, the river channel on the surface is dry for the majority of the time, unless there is heavy rainfall, in which case the underground channels fill up, causing water to flow overland as well. Far less erosion occurs in the surface river channel as a result of the infrequent channel flow.

Figure 16 The Bishopston River, South Gower: sink hole causing the river to go underground.

Example of concordant and discordant coastlines: Llyn Peninsula

- **Concordant**: the north coast of the Llyn Peninsula is a **concordant coastline**, where layers of different rock types run parallel to the coastline. The metamorphic rock erodes at the same rate so the coast has few headlands and bays.
- **Discordant**: by contrast, the coastline between the Trwyn Llanbedrog headland and Abersoch Bay is a **discordant coastline**. The headland consists of more resistant igneous rock (erodes slowly) and the neighbouring Abersoch Bay of less resistant mudstone and shale (erodes faster). The different erosion rates result in the formation of headlands and bays.

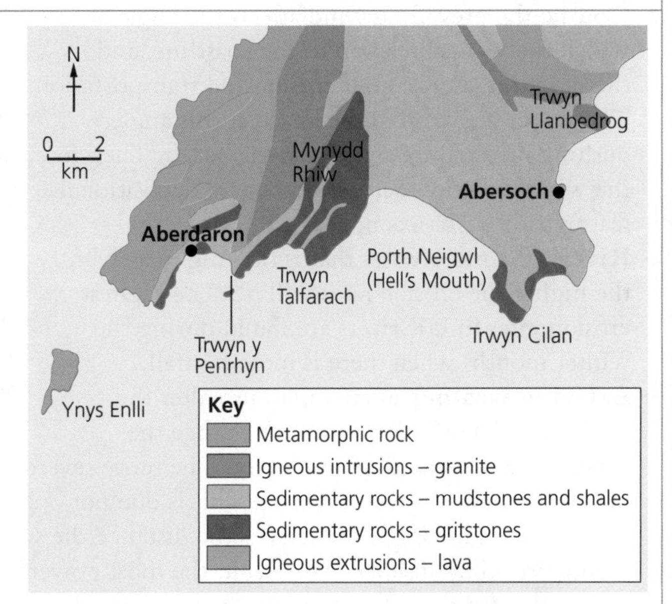

Figure 17 **The geology of the Llyn Peninsula.**

Exam tip

If you are asked to 'explain why' in a question about geology, make sure that you link the landform to the geological structure. You need to make it clear that the landform results **because** of the geology.

Revision activity

1 For an area of coastline that you have studied, complete the following bullet points:
- Name the feature
- Where is it located?
- What coastal processes occur here?
- What are the geological characteristics of this coastline?
- Which factor is the most dominant cause of the landforms found?

2 Repeat this exercise for a river feature that you have studied.

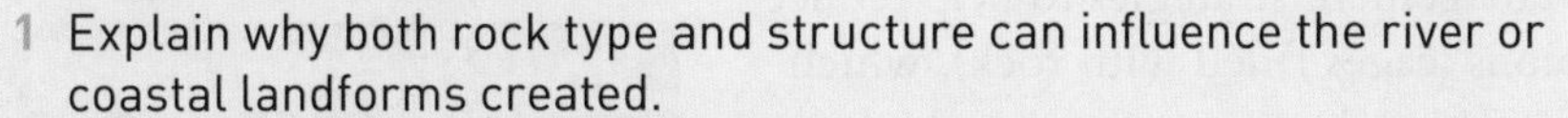

Now test yourself

TESTED

1 Explain why both rock type and structure can influence the river or coastal landforms created.
2 How can geology affect river landforms?
3 How can geology affect coastal landforms?
4 Is geology as important as erosion and deposition in the formation of coastal landforms?

Cavern A large underground cave which has been created due to enlargement of joints in carboniferous limestone

Sink hole A hole in the ground caused by a collapse of the surface layer, often found in carboniferous limestone areas where caverns are present

Concordant coastline Rocks are formed parallel to the sea so that erosion rates along the coastline are even

Discordant coastline Rocks are formed at right angles to the sea and so erosion rates vary along the coastline depending on rock type

Climate

Climate will affect the rate of change of river and coastal landforms:

- **Coasts**: the prevailing wind affects the angle at which the waves break on to the coastline, and therefore the direction of erosion and transportation. The waves break on to the beach at this angle, pushing material up and across the beach. Therefore, the wind direction determines where depositional features form (for example a spit).
- **Rivers**: the more water that is flowing in a river, the higher the erosion rates will be. The highest erosion rates in UK rivers are found during the winter months when there is more rainfall.
- **Extreme weather events** may also alter the landscape. A powerful storm can change the appearance of a coastline overnight. The more severe the storm, the more destructive waves it creates. This is due not only to the increased wind speed, but also to the fetch – the distance the wave has travelled before breaking onshore. As a result, the most powerful and destructive storms that hit the UK are usually from the south-west. This is due to the large distance of open water that the Atlantic Ocean provides, which increases the fetch of the waves.

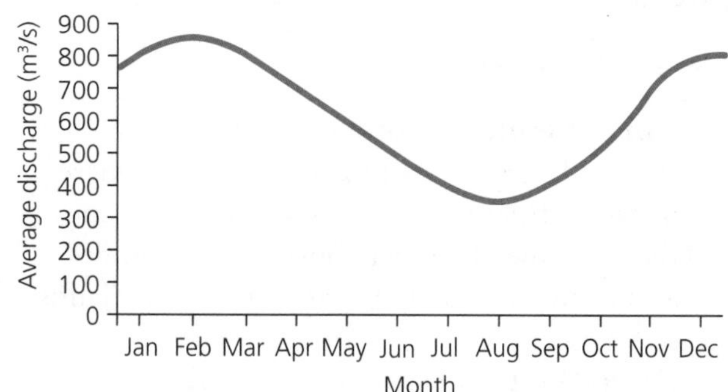

Figure 18 The river regime of the River Severn shows the variation in a river's discharge (the volume of water flowing through a river at any point) over a year.

Example of winter storms causing coastal damage: west Wales and East Sussex coastlines

In the spring of 2014, a series of storms hit the UK, causing severe erosion of the south and west coastlines. In Aberdaron, 30 cm of land was eroded from the cliffs, and at the Birling Gap on the East Sussex coast, 9 m^2 of cliff crumbled into the sea just metres from a cottage. This stretch of coastline suffered an estimated seven years' worth of erosion in just three months, due to the numerous storms.

Human activity

The impact of human activity on both river and coastal landscapes may be both intended and unintended:

- Intended human activity includes management strategies to reduce the impact of erosion of both river and coastal landforms.
- For rivers, an example is the management of meanders in a built-up area, where people are concerned about erosion on an outside bank affecting buildings and services. Management strategies to help reduce the amount of erosion include gabions (cages filled with rock), which absorb the power of the water, or reinforced concrete banks which resist the force of the water. These methods are used extensively on rivers running through built-up areas, for example the River Thames throughout central London.
- For coasts, an example is the management of beaches where the coastline is left exposed to erosion as the process of longshore drift moves large amounts of beach material along the coastline. As a result, human settlements close to the coast are vulnerable due to cliff collapse or flooding, for example St Bees Beach in Cumbria.

Now test yourself

1. Describe the impact of seasonal rainfall patterns on the discharge of a river.
2. Explain how extreme weather events can cause a change in the coastline.

TESTED

Example of intended human activity at the coast: St Bees Beach in Cumbria

The predominant south-westerly winds drive sediment in a north-easterly direction along the north Cumbrian coast from St Bees Head to the Scottish border, leaving the village of St Bees vulnerable to erosion. A series of **groynes** were built on the beach to restrict the movement of the sand and therefore maintain the beach for the important tourist trade and protect it from flooding and erosion.

Groynes A low wall or barrier on a beach built at right angles to the sea to restrict longshore drift

Figure 19 Groynes at St Bees Beach in Cumbria.

Example of impact of unintended human activity: Criccieth coastline

- Criccieth is on the south coast of Gwynedd.
- The cliffs are composed of easily eroded material known as glacial till.
- The beach is affected by longshore drift which moves the glacial till from west to east along the coastline.
- Groynes keep the material on the beach so that it continues to attract visitors and the income they bring.
- The beach also protects the coastline by absorbing the wave energy.
- To the east of the groynes there is an area of cliff that is very prone to collapse. This is likely to be due to the lack of beach material protecting it due to the groynes to the west.

Exam practice

1 Compare the influence of two different types of geological structures on the shape of coastlines. [8]

2 Explain why climate influences the rate of erosion in river landscapes. [4]

3 'Human intervention at the coastline will always have unintended consequences.' To what extent do you agree with this statement? [8]

Exam tip

When giving an example of the influence of human activity on either a river or a coastal landscape, make sure you explain the link between what it is that people are doing and the change in the natural environment.

Drainage basins of the UK

What physical processes affect stores and flows in drainage basins?

REVISED

Flows and stores of water

As water moves through a **drainage basin** it **flows** from one **store** to the next.

Relationship between drainage basin processes

The movement of water through the drainage basin can be seen in the diagram. The speed at which water moves through drainage basins can vary due to:

- **The type and quantity of rainfall**: rainfall moves more quickly through a drainage basin in heavy rainstorms than in light drizzle. Raindrops are larger and falling in a shorter period of time, so less infiltration and more surface runoff will occur.
- **The type and quantity of vegetation cover**: more **interception** occurs in a woodland compared to a meadow.
- **The size and shape of the drainage basin**: round drainage basins lead to a faster movement of water into the river than elongated ones. Larger basins have a larger discharge as they drain from a larger surface area of land.
- **The steepness of slopes**: steeper slopes mean greater surface runoff and less infiltration.
- **The geology and soil type within the drainage basin**: impermeable soil or rocks lead to less infiltration or groundwater flow and more surface runoff.

Drainage basin An area of land drained by a river and its tributaries

Flow The movement of water

Store A place where water is stationary within the water cycle

Interception When rainfall does not reach the ground as it is blocked by trees, buildings and so on

Infiltration The movement of water into the soil

Throughflow The flow of water through the soil

Overland flow The flow of water across the ground surface

Groundwater flow The flow of water through rocks

Transpiration Water given off by plants

Stem flow Movement of water that has been intercepted down the stem or trunk of a plant

Percolation The movement of water from the soil into the bedrock

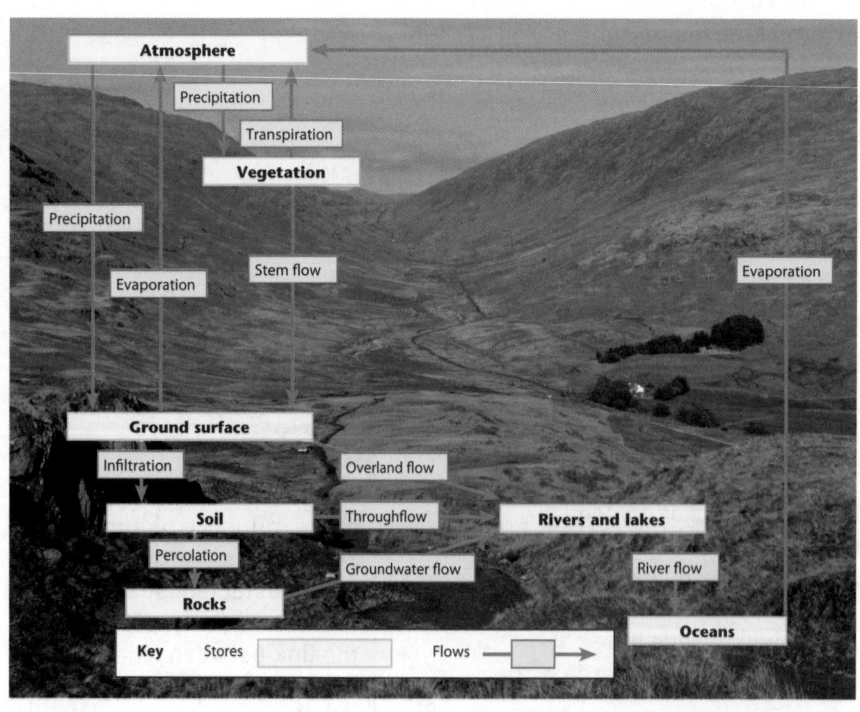

Figure 20 Stores and flows of water in a natural drainage basin.

Now test yourself

TESTED

Make a list of the characteristics of a river basin that would be likely to lead to a flood.

Exam practice

1 Name two stores and two flows in a drainage basin. [4]
2 Explain the impact on the rest of the drainage basin system of cutting down a large area of trees. [6]

Exam tip

When asked to discuss the interrelationships between drainage basin processes, you will need to explain why a 'store' in one part of the basin may affect the 'flow' in another part.

Why do rivers in the UK flood?

REVISED

Climate, vegetation and geology (physical factors) may affect a river's discharge and lead to flooding. Urbanisation (human factors) may also cause flooding.

Why do rivers flood?

Climate

Increased amounts of rainfall will increase the chance of a river flooding. This may be due to either:

- Seasonal rainfall: continuous rainfall causes the ground to become saturated, leading to more overland flow and therefore higher river levels
- A storm event: when a heavy storm brings a high volume of rainfall in a short period of time this causes a sudden rise in river levels that can lead to flash floods

Vegetation

- Different types of vegetation intercept different amounts of rainfall, which influences how rapidly water moves through the drainage basin to reach the river channel. For example, broad-leaved trees intercept more rainfall than grassland and reduce the speed at which it reaches the ground. In addition, their roots are deeper and cover a larger area, so taking more of the infiltrated water out of the soil
- Removal of vegetation: if trees are removed from a drainage basin then water will reach the river channel much more quickly due to quicker saturation of the soil

Geology

- Porous rocks have large spaces within the rock which allow water to pass through. This reduces the flood risk due to increased groundwater flow
- Impermeable rocks have very few spaces within the rock. Little water passes through, creating more overland flow and a higher flood risk
- A rock may not be porous but can be well jointed, allowing water to pass through these lines of weakness. For example, carboniferous limestone

Urbanisation

The expansion of towns and cities leads to the ground being covered with impermeable surfaces such as tarmac, which reduces the amount of water that is infiltrated. This causes more overland flow and increases the chance of flooding

Exam tip

When asked to 'explain why' a factor increases the risk of flooding, you must state what it is about the factor that would lead the river to flood.

Hydrographs

A **hydrograph** is a good way to see how a river will react to a storm event. It shows the total rainfall amount represented by a bar graph and the river discharge as a line graph. These are plotted against time. The shape of the hydrograph indicates whether a river is likely to flood after a storm event.

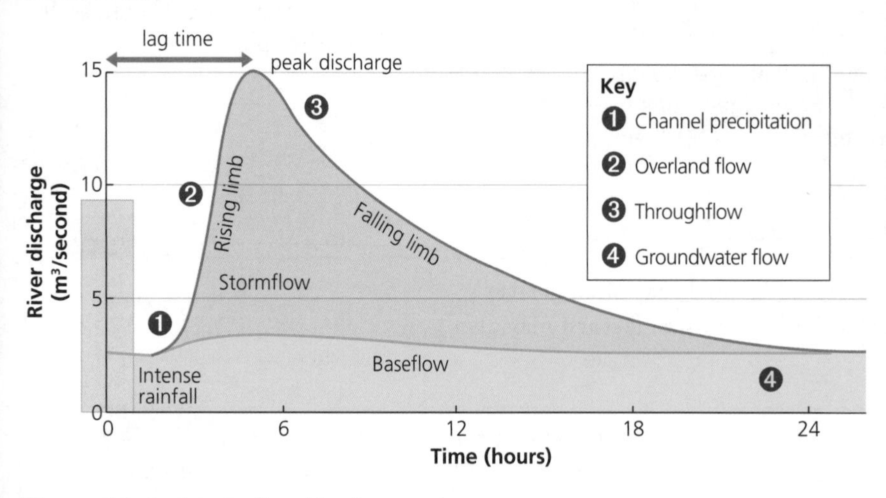

Figure 21 A simple flood hydrograph.

Hydrograph A line graph used to display the discharge of a river over a period of time

Lag time The time between the peak rainfall and peak discharge in a river

Rising limb The part of a hydrograph where the discharge of a river is increasing after a rainfall event

Factor	Impact
Urbanisation	Decreases **lag time**, steep **rising limb**, high peak
Porous rocks	Increases lag time, gentle rising limb, lower peak
Impermeable rocks	Decreases lag time, steep rising limb, high peak
Broad-leaved trees	Increases lag time, gentle rising limb, lower peak

Example of causes and effects of flooding: Somerset Levels

The winter of 2013–14 resulted in unprecedented amounts of flooding on the Somerset Levels.

Causes	Effects
Physical • Large area of flat land • Two major rivers run through the area – the Tone and the Parrett • Prolonged heavy rainfall causing saturated ground • Silting up of the river channel therefore reducing capacity • High tidal range **Human** • Reduction in the frequency at which the rivers were dredged • Building on the floodplain	• Over 80 roads were blocked • Children were not able to get to school or people to work • Villages were cut off • 600 homes were affected and people evacuated • Due to lack of grazing land, farmers lost around £10m • Half of businesses in Somerset lost money • Cost of £19m to local government and emergency services

Now test yourself

1. What is meant by the term 'annual regime'?
2. Describe how climate can affect the discharge of a river.
3. List the different ways in which people can increase the chances of a river flooding.
4. Explain the different parts of a hydrograph and the factors that can influence them.

TESTED

Exam practice

1. List three factors which may affect flooding in the UK. [3]
2. Describe how porous rocks will influence the shape of a hydrograph. [4]
3. 'Urbanisation is the most important factor in causing flooding in the UK today.' To what extent do you think this is true? [8]

What are the management approaches to flooding in the UK?

REVISED

Strategies for river channel and drainage basin management

Hard engineering strategies	Soft engineering strategies	Land-use zoning
Involve constructing defences to control natural processes	Work with the environment rather than trying to control it	Planning what the land is used for within a river basin so that less valuable land is closer to the river, for example grazing land and playing fields
They are often large scale, expensive and relatively effective	Often cheaper than hard engineering and have less impact on the environment, but can be less effective once the river has flooded	Housing and key services are usually put on higher ground away from the river to lessen the chance of these being flooded
Examples include: dams, artificial levees/embankments, artificial river channels (channelisation), gabions, dredging the river channel, creating a flood relief channel	Examples include: restrict building on flood plains, afforestation, ecological flooding, warning systems	This method is not always straightforward to carry out, as in many towns and cities, housing and key services were built close to rivers many years ago

Conflicting views

Different groups of people often have very different views when considering the best way to manage a floodplain or when considering future development on a floodplain.

Example of floodplain management: Somerset Levels

After serious flooding in early 2014, there has been much debate about flood prevention in the future. It was reported that a lack of dredging of the river channels was a major cause of the flooding and many people thought dredging would prevent flooding in the future. Others disagreed:

- The Somerset Wildlife Trust: 'We must allow the rivers to flood naturally. Some of the wetland habitats are unique.'
- The Royal Society for the Protection of Birds (RSPB): 'Dredging has an impact on the habitat in and around the river.'

Floodplain development

An increase in the demand for housing is leading to more pressure to build on floodplains. Houses built on floodplains are at a higher risk of flooding.

Example of floodplain development conflict

In 2013, local councils in England and Wales granted 87 planning applications for new housing developments on floodplains, all of which caused conflict in their local communities.

- Insurance broker: 'Over 80 new developments to be built on floodplains is a ridiculous amount, there should be none! If people don't want their homes to be flooded, don't build on floodplains!'
- Young family on low income: 'We need more homes to be built so that the price of houses falls low enough for us to be able to afford them.'

Revision activity

Make a table of the advantages and disadvantages of dredging as a method of floodplain management.

Exam practice

Evaluate the effectiveness of soft engineering as a strategy for managing UK floodplains in the future. [8]

Theme 2 Rural–Urban Links

The urban–rural continuum

How are urban and rural areas linked?

REVISED

The human landscape of the UK is as diverse as the physical environment. **Rural** areas are characterised by sparse or low population densities whereas **urban** areas are busy built-up environments with a higher **population density**. Both of these types of environments are linked and influenced by each other.

The map shows the **location** of significant areas of **population** in the UK.

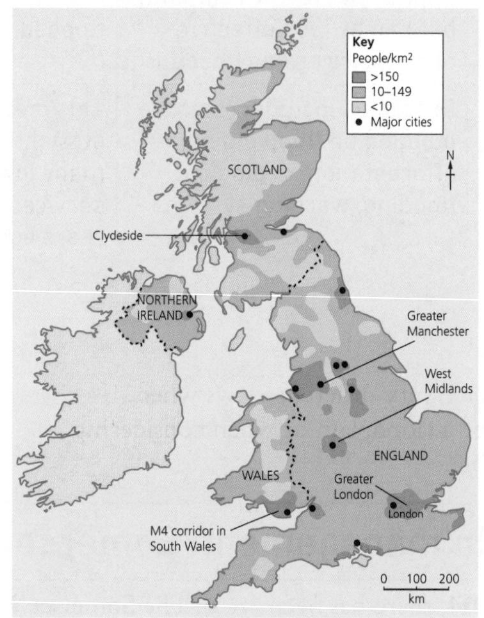

Figure 1 **Population distribution in the UK.**

Urban–rural continuum

Often, the concept of a settlement hierarchy is difficult to apply because of the changing nature of many villages and towns. An **urban-rural continuum** allows us to consider the two extremes and then place all other types of settlements in between. We can then describe a particular settlement as 'more rural' or 'more urban' in character. The diagram below illustrates this.

Isolated farmhouse → Hamlet → Remote village → Village → Suburbanised village → Town → City → Mega-city → Large urban area

As a settlement moves along the continuum from rural to urban it develops more functions and provides more services. For example, a small village is likely to provide the following:

- a post office
- a small shop
- a bus stop
- a pub
- a church

whereas a city is likely to provide:

- shopping centres with many chain stores
- entertainment centres such as theatres and cinemas
- a variety of restaurants and bars
- a range of medical facilities including a hospital.

Rural An area of countryside characterised by wide open spaces

Urban A built-up environment where a lot of people live

Population density The average number of people per square kilometre. Usually expressed as densely or sparsely populated

Location A particular place or position

Population The number of people living in an area

Urban–rural continuum A continuum along which all settlements are placed

Exam tip

It is important that you are able to describe the location of areas of high population density in Wales and the rest of the UK.

The services that are actually provided in any city will depend on the individual city and its location in relation to other urban areas.

Now test yourself

TESTED

1 Make a table like the one below to list locations of high and low population density in the UK.

Areas of high population density	Areas of low population density

2 Pick two areas of high population and describe their location.
3 Give reasons as to why the high-density populations are located there.

Urban sphere of influence

The function and services of a city serve not only the residents of the city but also the smaller settlements that may surround it. This is referred to as a city's **sphere of influence** and its strength is mainly due to:

- The infrastructure and transport links between the urban and rural areas: the more widespread and efficient the road and rail networks are, the wider the sphere of influence.
- The distance from the urban area: the influence is greater the closer a rural settlement is to the city.
- The size of the urban area: the sphere of influence will be greater for a large city than a small town.

Sphere of influence A region within which an urban area provides an important economic and social influence

Example of a city's sphere of influence: Liverpool and Glyn Ceiriog

- Liverpool is a large city on the north-west coast of England, close to the north-eastern border of Wales.
- It is the closest, largest urban area to north-east Wales, with a population of 467,250.
- Its sphere of influence over the region includes Glyn Ceiriog, a small village of 800 people in north-east Wales.

The services available in Glyn Ceiriog include:

- two shops (including a post office)
- a pharmacy
- a GP's surgery
- two places of worship
- a hotel
- a pub.

It is likely that the residents will need to go outside the settlement for many services, for example to buy anything that the local stores could not provide. Wrexham and Chester offer a variety of retail stores, but for a larger choice Liverpool city centre has a range of major shopping areas.

In rural areas, access to healthcare is also restricted and for specialist health services the residents of Glyn Ceiriog would need to travel to Wrexham and Liverpool.

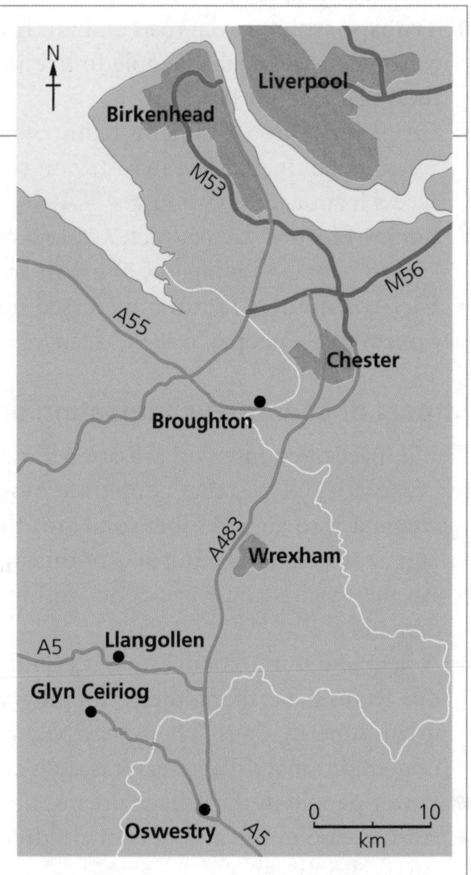

Figure 2 Urban and rural places on the border between England and Wales.

Now test yourself

TESTED

1. Explain why a large urban area would have a sphere of influence over a neighbouring rural area.
2. Why does the sphere of influence lessen with distance from the urban area?
3. Make a list of the retail and healthcare services that an urban centre could provide to rural areas.
4. Make a list of factors other than retail and healthcare for which urban areas could have an influence over rural communities.

Revision activity

Draw a spider diagram to illustrate the influencing factors that a large urban area that you have studied has over its neighbouring rural community.

Counter-urbanisation

Since the 1980s, the UK has been experiencing **counter-urbanisation**. This process involves the movement of people, such as young families or retired people, away from the inner cities to either:

- the urban–rural fringe
- suburbanised villages in rural areas that are easily accessible to the city, or
- remote rural areas.

Counter-urbanisation The movement of people out of towns and cities to rural areas

Reasons for counter-urbanisation

- **Housing**: the type and style of homes people want are more available and affordable in rural areas.
- **Family status**: with increased income or family size, people may look for larger properties in rural locations.
- **Transport**: improved road and rail links, together with the increase in car ownership, enable people to live in a different location and travel to their place of work.
- **Employment**: a decline of industry being located in central urban areas and an increase of industry on the rural–urban fringe improves access from rural locations.
- **Social factors**: factors such as a low crime rate and good schools often lead to people moving out of urban areas.
- **Environmental factors**: increased noise and air pollution in urban centres cause people to seek 'cleaner' environments to live in.

Now test yourself

1. What do you understand by the term 'remote rural area'?
2. Why does counter-urbanisation take place?
3. What impact does counter-urbanisation have on rural settlements?
4. For a rural settlement that you have studied, describe the physical impacts of counter-urbanisation.

TESTED

Impact of counter-urbanisation on rural settlements

The impact of counter-urbanisation varies greatly and depends on the type of rural area that people are moving to. For example, if the movement is to an accessible rural area then a 'dormitory' or 'commuter' village may result. The impacts of this may include:

- An increase in house prices due to higher demand.
- A decrease in traditional services (village shops) due to residents doing their shopping at larger urban supermarkets. There may be an increase in non-traditional village services such as a crèche.
- Fewer people in the village during the daytime.
- An increase in the numbers of children attending rural schools.
- Increased amounts of traffic and associated pollution on rural roads.
- Loss of villager 'identity' as the majority of residents do not work in the village.

Revision activity

For a rural settlement that you have studied, create a fact file to explain the reasons for and describe the consequences of people moving into the village. Try to include:

- location and population size
- what attracts people to the village
- changes to buildings
- changes to the occupations of the residents
- changes to the local school
- changes to the services the village offers
- issues related to transport.

Patterns of commuting

Commuting significant distances to work is common for many people. As cities such as London and Cardiff offer many more job opportunities than the rural areas that surround them, many people commute to the city to work. People often choose to live in a cheaper rural area and travel a longer distance to work rather than pay the higher house prices that are found in cities. For others, the internet and mobile phones remove the necessity to commute as they allow them to work from home. The table below shows the factors that both encourage and discourage commuting.

Exam tip

When describing issues that arise from commuting, refer to specific information for an urban area that you have studied so you can show your depth of knowledge.

Factors leading to increased commuting	Factors leading to decreased commuting
Cities have more job opportunities than rural areas	Rapid growth of the internet and email removes the need to be in the same physical office as co-workers
People choose to live in rural housing, which is often cheaper than inflated city house prices	Increased coverage and quality of mobile phone network providers enable people to stay constantly in touch with co-workers
Improvements in road and rail links have cut journey times	Rapid growth of broadband has led to many companies encouraging employees to work from home
Improvements in car safety and comfort have encouraged more people to travel longer distances	

Example of transport issues arising out of counter-urbanisation: Cardiff

Cardiff is located in south-east Wales. As Wales' largest city and its capital, Cardiff attracts many industries and has a wide sphere of influence. This influence is strongest in south-east Wales. People travel to Cardiff to work, shop and attend sporting events. The pattern of commuting in south-east Wales involves:

- 78,000 people travelling into Cardiff every weekday.
- 33,900 people travelling out of Cardiff every weekday.
- The Vale of Glamorgan is the highest source and destination region for commuters in and out of Cardiff.
- The M4 motorway together with main access roads into Cardiff have been improved over the past twenty years.

With around 112,000 people travelling either to or from Cardiff every weekday, issues arise from the large number of people on the move:

- The main routes into and around Cardiff (A470, M4 and A48) become very congested during the rush hour, leading to long delays and increased journey times.
- Increased amounts of air and noise pollution affect Cardiff and the roads on its periphery where traffic congestion builds up.
- Roads within the city become congested and dangerous to cyclists and walkers.
- Trains arriving in Cardiff during the rush hour are packed with commuters having standing room only.

Exam practice

1 Give the definition of the term 'counter-urbanisation'. [2]
2 Suggest and describe two examples of economic impacts of counter-urbanisation. [4]
3 For an example that you have studied, describe the pattern of commuting. [4]
4 Explain the issues created by commuting in your named example. [6]

Revision activity

For an urban area that you have studied, use a spider diagram to illustrate the issues that have resulted from commuters travelling to that town or city.

How are rural areas changing?

REVISED

Impacts of urban spheres of influence and technological change on service provision

We have already seen how urban spheres of influence have an impact on retailing and health services in rural communities, but these are just some of the changes that are happening. Coupled with changes in technology (improved broadband coverage and mobile phone reception), many rural areas are changing in both size and character. Some of the changes include:

- Reduction and/or change in employment opportunities in the rural area, for example a reduction in traditional **primary jobs** and an increase in **tertiary jobs**.
- Closure of rural banks and post offices, centralising services in urban areas in addition to an increase in online banking.
- Increase in house prices in accessible rural areas within the '**commuter belt**'.
- An increasing number of second homes in picturesque parts of the countryside, for example Pembrokeshire in Wales and the Lake District in England.
- Local people being unable to afford to buy a home due to the increase in prices, forcing them to move away from the village.
- Closure of village shops. With the increase of online grocery shopping and home delivery services by many supermarkets, village shops are under even more pressure to close.
- Reduction or removal of bus services due to few people using them.

Primary jobs Jobs that involve getting raw materials from the environment, for example fishing, mining and farming

Tertiary jobs Jobs that provide a service, for example teaching, medical and retail

Commuter belt The area around a town or city where people travel to work in the urban area

Depopulation The reduction in the population of an area

Deprivation The lack of key features that are regarded as necessary for a reasonable standard of living

Cycle of deprivation The cycle where a family living in poverty is unable to improve its lifestyle due to the negative factors of low income, poor housing and education, which keeps it in a state of poverty

Rural poverty, depopulation and deprivation

Remote rural areas have experienced a lot of negative change in recent years. This has led to **depopulation** in many villages and a high level of poverty and **deprivation** experienced by the villagers who remain. Deprivation in rural areas is usually characterised by a lack of public transport, healthcare and education services. It is often referred to as the **cycle of deprivation** due to the nature of the impacts that are created becoming a cause for further depopulation and deprivation.

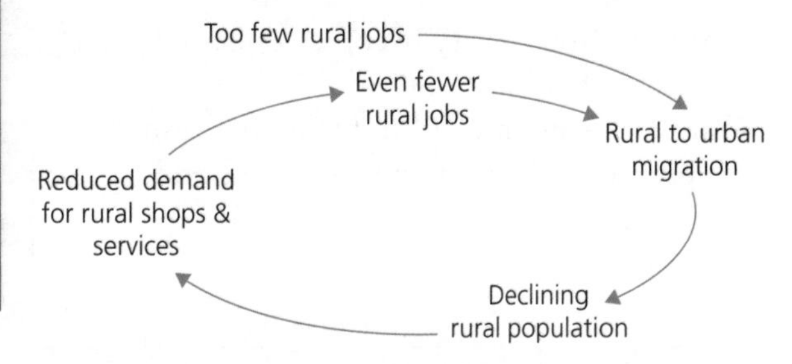

Figure 3 The cycle of deprivation.

Now test yourself

TESTED

1. What do you understand by the term 'technological change'?
2. How has technological change impacted on service provision in settlements?
3. Explain why it is difficult to break out of a cycle of deprivation.

Revision activity

Draw a double-bubble diagram like the one below and decide which of the changes are caused by technological change, which are caused by the urban sphere of influence and which are the result of both. Colour code your diagram.

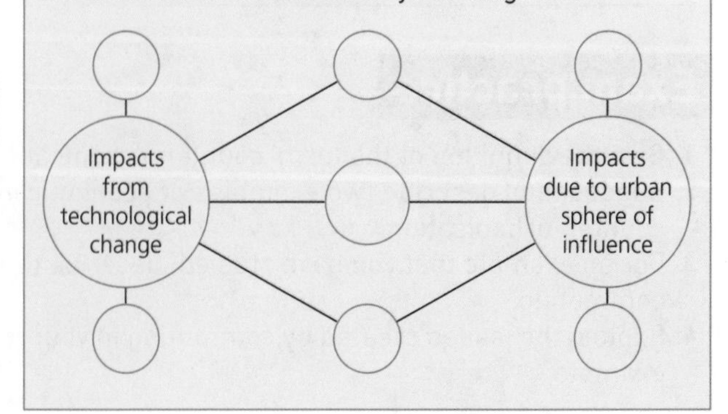

Creating sustainable rural communities

A **sustainable community** is the aim of all local planners. Proposals for new housing or roads, and decisions on what transport, education and health services to provide, need to be considered in terms of the impacts on local people and the environment. In a rural setting, providing access to services can be challenging due to the smaller number of people using those services, and the resulting higher cost per person.

Sustainable community
A community which is able to support the needs of all its residents with minimal environmental impacts

Strategies that could be used to create a sustainable rural community are:

- **Reliability and frequency of transport**: ensuring that public transport is available when the rural community needs it.
- **Availability of jobs**: securing investment from companies to develop jobs based in the rural area.
- **Internet connections**: ensuring the availability of fast and reliable broadband.
- **Education**: ensuring that village schools remain open and secondary schools offer a wide range of subjects taught by specialist staff.
- **Healthcare**: ensuring access to all aspects of healthcare with a provision of transport if needed.
- **Village services**: encouraging village shops, pubs and post offices to remain open for residents to use.
- **Green technologies**: promoting the use of renewable energy within the rural area.

Implementing these strategies will depend on their cost but using them as a target will help to make communities more sustainable.

Exam tip

Link the strategy employed to how it makes the community sustainable. For example, if a new bus service was to go through a remote rural community three times a day, this would make the community more sustainable because it is enabling people to continue living there while they may work in a nearby urban area, therefore reducing the number of cars on the road and increasing sustainability.

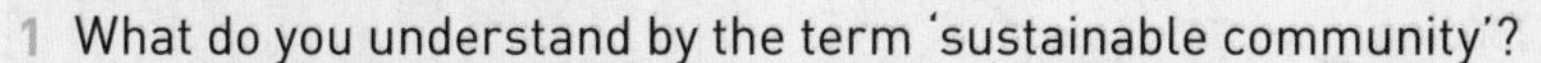

Now test yourself

TESTED

1. What do you understand by the term 'sustainable community'?
2. Suggest strategies that could be taken to make rural communities more sustainable.

Exam practice

1. What do you understand by the phrase 'rural poverty and deprivation'? [2]
2. Explain why some rural communities suffer from rural deprivation. [4]
3. Discuss the challenges that are faced in creating sustainable rural communities. [8]

Revision activity

For a rural community that you have studied, pick one of the above strategies and research how that rural area is addressing the issue. Then write a proposal to the local planning office to show how it could be made more sustainable.

Population and urban change in the UK

What are the causes and consequences of population change?

REVISED

Changing rate of population

The population in the UK is always changing. In 2014, the UK population was 64.8 million while in Wales the population was 3.12 million. Between 2004 and 2014, the population in the UK grew, although the rate of change varied in different areas.

Growth	Wales	UK
Highest growth rate	Cardiff +1.7 per cent change (increase)	Tower Hamlets +34.5 per cent (increase)
Lowest growth rate	Ceredigion –0.5 per cent change (decrease)	Redcar & Cleveland –2.6 per cent (decrease)

Natural population change The change in population from births and deaths only

Migration The movement of people from one place to another

Social factors Factors that relate to people's health, lifestyle and community

While the population is increasing in some areas of the UK, elsewhere it is decreasing. A change in population can be the result of **natural population change** or **migration**.

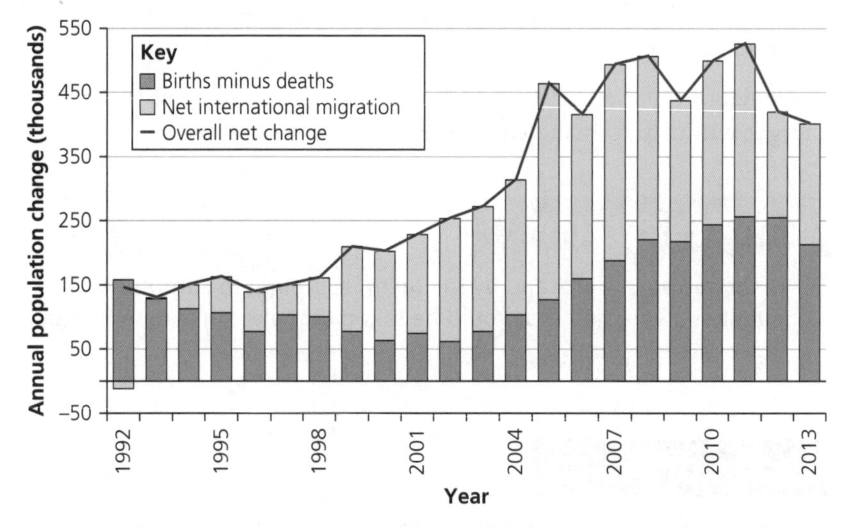

Figure 4 Annual UK population change, 1992–2013.

Factors affecting population change

Social factors:

- **Healthcare**: as the UK has a freely accessible health service (NHS) residents are able to access the care they need, leading to a longer life expectancy and low **infant mortality rates**.
- **Marriage**: people are marrying later in the UK, which may affect when they start a family.
- **Culture**: it is widely acceptable in the UK for women to delay starting families until their 30s, and this may reduce the total number of children they have.
- **Ageing population**: with a greater proportion of people in the UK being past their child-bearing age, this will naturally lead to a lower **birth rate**.

Economic factors:

- **Cost of raising a family**: this has increased in the UK and may discourage people from starting a family.
- **Maternity pay**: the longer period of statutory maternity pay may encourage more births.
- **Career**: many women choose to continue with their careers and increase their incomes rather than start families.

Infant mortality rate Number of deaths of children under one year of age occurring per 1000 live births in an area per year

Ageing population Country which has a high proportion of people aged over 65

Birth rate The number of babies born in an area per 1000 of population

Economic factors Factors that relate to cost and finance

Political factors:

- **Contraception**: contraception is widely available in the UK and education is provided so that unwanted pregnancies may be restricted.
- **Maternity/paternity rights**: changes to maternity and paternity leave and pay for parents may encourage people to have more children.
- **Migration**: both into and out of the UK. In 2015, 333,000 more people moved to the UK than moved out, mainly from Commonwealth countries and the European Union (EU).
- **Vaccination programmes**: the UK has a programme of vaccinating children, which reduces the **death rate**.

Migration into and within the UK

In 2015, 630,000 people moved into the UK. The UK is a country of choice destination for many people, but this is only half the migration story. Many people also migrate within the UK. This may be rural to urban or urban to rural migration, within the same urban area or to a different part of the UK altogether.

Reasons for people moving into the UK	Reasons for people moving within the UK
Availability of jobs	Cost of housing: move to an area where they can afford to buy a property
Stable political system	Change of lifestyle: people may retire to a rural location
Good health service	In search of work: people move for work or in order to further their career
Better rates of pay; therefore a higher income	The need to live close to your place of work is increasingly becoming less important
Good education system	Locate near family for care needs
Already established network of family or people of similar ethnic or cultural origin	

The impacts of both types of migration may include:

- an increase in the number of young adults who are able to work and pay taxes
- an increase in the birth rate due to migrants having children
- an increase in the number of languages that are spoken in the UK
- a strain on schools due to the number of non-English-speaking pupils that they have
- many low-paid, unskilled jobs that UK citizens may not wish to do are filled
- house prices in more desirable areas increase
- diversity of culture means the UK has a greater range of foods, restaurants and so on.

Revision activity

Using the double bubble that you have drawn for the Now test yourself activity, construct a table similar to the one below to help you revise impacts of migration.

Factors	Inward migration	Domestic migration
Social impacts		
Demographic impacts		
Economic migration		

Now test yourself

Look at the graph showing the annual UK population change.

1. Describe the trend of natural change seen in the UK.
2. Describe the trend for migration in the UK.
3. Explain why the way that population changes is altering in the UK.

TESTED

Political factors Factors that relate to decisions made by government, either national or local

Death rate The number of people dying in an area per 1000 of population

Now test yourself

1. Draw a double-bubble diagram like the one on page 32 to show the impacts of inward migration and domestic migration.
2. Colour code these impacts to illustrate which are social, demographic (impacts that affect the structure of the population of an area) and economic impacts.

TESTED

Challenges of an ageing UK population

Due to the low birth and death rates in the UK, an increasing proportion of the population is aged over 65. As the percentage of this group of people is increasing compared to the rest of the population, the UK is said to have an ageing population. In 1995, there were less than 9 million people aged over 65, whereas in 2030 it is projected that there will be 13 million. Such a large proportion of elderly people in a population presents economic, health and social challenges which need to be addressed.

Economic challenges	Health challenges	Social challenges
A reduced number of economically active people in the UK to pay taxes	Increase in health issues as people live longer	Older people have a wealth of knowledge and skills that will be lost if not passed on to the younger generation
More money needed to pay for state pensions	Large increase in care services required to look after people in the community	Increasingly, working-age people are caring for their children and their elderly parents
More people are dependent on the state		Increase in the number of elderly people living on their own, which may lead to feelings of isolation and also decrease the amount of housing available

Revision activity

- On an A5 revision card, draw a two-column table.
- List the challenges in the left-hand column.
- In the right-hand column explain what impact these challenges may have on the UK.
- Then colour code the challenges: red for positive and blue for negative.
- Recommend possible solutions to the challenges that you have listed.

Exam tip

Ensure that you are able to identify the difference between economic, health and social challenges as the exam question may just ask about a specific type of challenge.

The need for new housing

Increased immigration, longer life expectancy and an increase in single-occupancy households all lead to an increase in the number of homes required. The government set a target to build a million new homes by 2020. In August 2016, the government admitted that it was going to miss this target by 266,000 homes. In missing this target the price of existing housing is likely to increase due to demand and supply problems. The demand for housing will vary across Wales and the UK, and is highest usually where the economy is strongest. In Wales, the south-east region has one of the highest demands for housing due to the jobs that the capital city of Cardiff attracts. In the UK as a whole, it is the south-east of England that attracts most people to move there for employment, therefore creating the highest demand for houses. The graph in Figure 5 shows how house prices have changed since 2007.

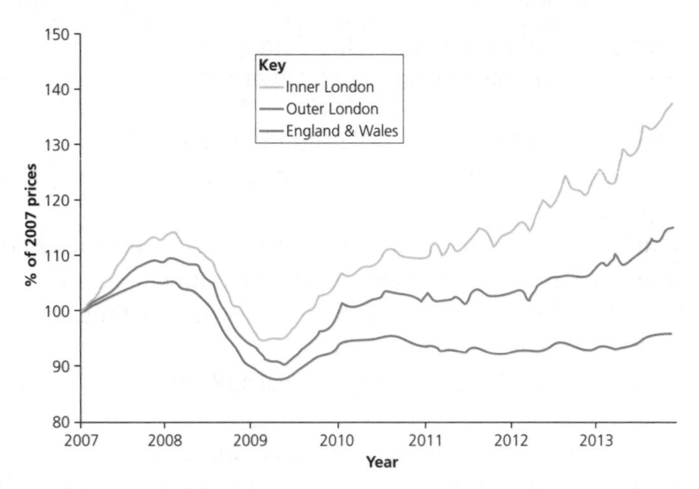

Figure 5 The change in house prices in England and Wales since 2007.

Exam tip

When a question asks you to 'describe a trend', you must focus your answer on describing how the trend is changing, giving the highest and lowest points and an overall statement about the trend rather than explaining why this trend is occurring.

Now test yourself

TESTED

Look at the graph in Figure 5:

1 Describe the trend in house prices between 2007 and 2013.
2 Explain the variations that can be seen between London and the whole of England and Wales.
3 Explain why building new housing developments may lead to a decrease in house prices.

Exam practice

1 List three different examples of migration that the UK is experiencing. [3]
2 Explain two housing issues that the UK is currently experiencing. [4]
3 Evaluate the various options available to alleviate the current housing crisis in the UK. [8]

What are the challenges facing UK towns and cities?

REVISED

Challenges of creating urban sustainable communities

The guiding principle of building sustainable communities is to ensure that any new developments (roads, housing, industry, communications and so on) benefit both the community and the environment in the short and long term. The characteristics of a sustainable community are illustrated in Egan's wheel in Figure 6.

It is not always as simple as building a brand new community. Usually an existing urban area is being added to or being renewed, making it difficult to achieve the goals set out above. The challenges include:

- Building on greenfield land is often necessary to increase the number of homes.
- Community differences: not everyone in the community wants the same thing.
- Existing community: people often do not want change in their local area.
- **Eco-housing** often costs more to build and may not be affordable.

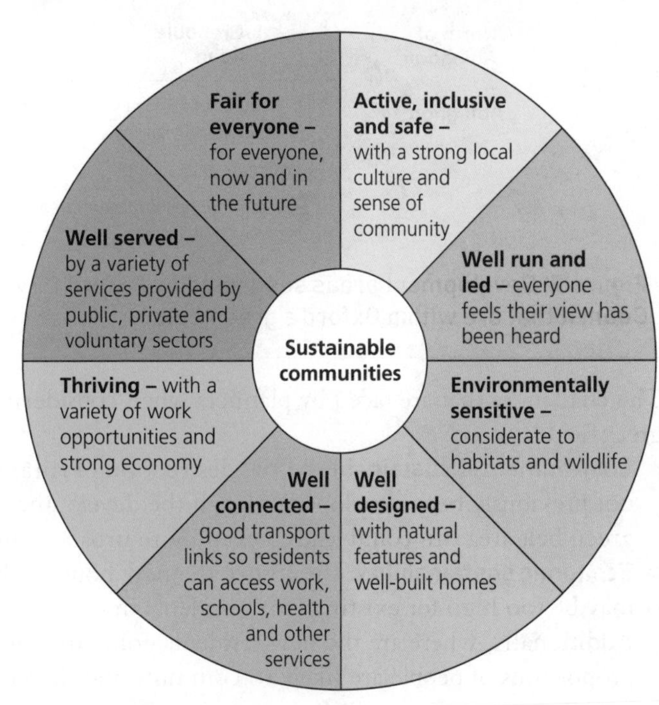

Figure 6 Egan's wheel.

How the planning authorities overcome these challenges will vary from place to place.

Eco-housing Houses that are built to make them environmentally sustainable

Example of a greenfield site: Oxford's green belt

Green belts were put in place around UK cities as early as 1935 to try to stop the expansion of towns and cities. They occupy thirteen per cent of total land area in England but are coming under increasing pressure to be built on because of the high demand for new homes. An example of **greenfield sites** that have been identified for potential development is on the green belt surrounding Oxford. The map in Figure 7 shows possible locations.

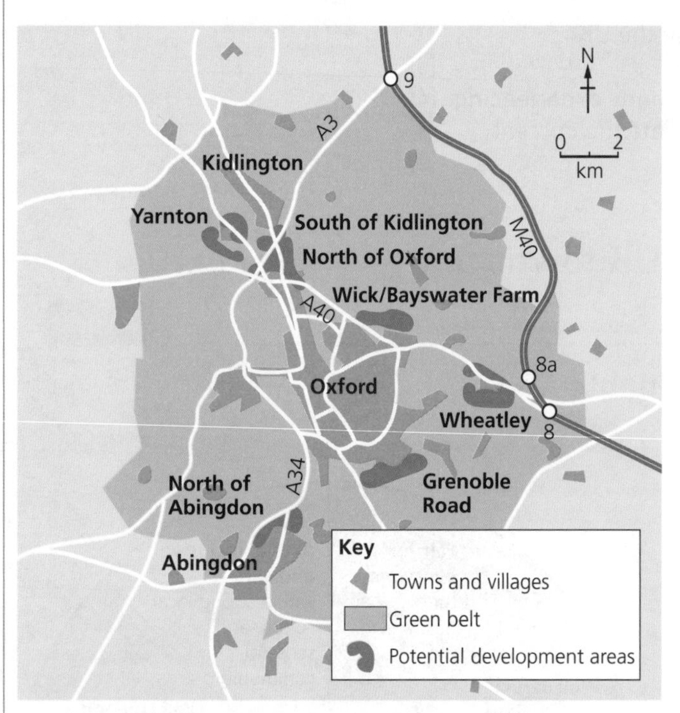

Figure 7 Development areas suggested by Oxford City Council that are within Oxford's green belt.

Greenfield site An area of land that has not been used before for building

Environmental sustainability Improvements in the standard of living that do not cause long-term damage to the environment

Economic sustainability Development that ensures everyone has the right to economic improvement in the long term

Social sustainability Development that is inclusive and ensures an improvement in the standard of living for all

The challenges that are faced by planners when considering building on greenfield land are:

- **Environmental sustainability**: the destruction of rural land that has not previously been developed. Also, if the development is within a green belt area this could lead to even more urban sprawl.
- **Economic sustainability**: the cost of the new housing developments may be too high for existing local residents in the rural area to afford. Additionally, where are the new residents going to work? High proportions of people are likely to commute and therefore leave the new developments empty during working hours, which will not help rural services.
- **Social sustainability**: new greenfield developments may encourage urban residents to move to the countryside and therefore change the way of life of the rural area. In addition to this, local services such as schools and GPs' surgeries will be put under pressure.

Example of a brownfield site: Ipswich Waterfront Development

Building on **brownfield sites** has become very fashionable in the past 30 years for two main reasons. First, as these sites have previously been built on (unlike greenfield sites), they do not destroy areas of countryside. Second, many people now like to move back to revitalised inner-city areas – a process known as **re-urbanisation**. Many cities with old industrial waterside locations are regenerating those areas into highly desirable living and entertainment locations. An example of such a development is in Ipswich with its Waterfront Development. The key features of this are:

- The site was formerly an industrial dockland area with warehouses and factories.
- It had been left derelict since the 1970s.
- Warehouses have been refurbished into shops, entertainment locations and housing.
- The land is more expensive than greenfield land to develop due to having to clean it up before building could take place.
- Remains of archaeological value needed to be conserved at a cost of £1.2 million per hectare.
- A former derelict part of the city has been transformed into a vibrant location.

Brownfield site An area for redevelopment that has previously been built on

Re-urbanisation People moving from the countryside back to urban areas

Now test yourself

TESTED

Draw a table similar to the one below and fill in as many challenges as you can think of for the different approaches to creating a sustainable urban area.

Brownfield sites	Greenfield sites

Exam practice

1 Describe any two Egan's wheel parts. [4]

2 'Achieving a sustainable community is equally as difficult on a greenfield site as on a brownfield site.' Discuss why this statement may be true. [8]

Exam tip

For any examples of developments that you learn, ensure that you can describe where they are located or even draw a sketch map.

How and why is retailing changing in the UK?

REVISED

The traditional shopping hierarchy of the **central business district (CBD)**, with its large shopping centre or mall, the local high street offering a smaller range of shops and the corner shop selling convenience goods, has for some time been under threat from the rise of the out-of-town retail parks and internet shopping.

Central business district (CBD) The main shopping and service area in a city

Factors leading to a change in retailing

Economic factors	Cultural factors	Technological factors
Increase in the number of home delivery firms, making delivering goods cheaper	Car-dependent society	Development of wide coverage of high-speed broadband
Congestion in city centres	Habit of bulk buying and weekly or monthly shops	Sophisticated websites which can show you your goods from any angle before you buy
Large areas of free parking in out-of-town retail parks		Rise of many retailers that are only online
High city-centre parking costs		Internet banking
Wages being paid monthly rather than weekly		

Costs and benefits of out-of-town shopping centres

Benefits	Costs
Large free parking areas for customers	Attracts shoppers away from city centres, which could cause their decline
Quick and easy access for customers and deliveries due to being located near major road junctions	Can cause congestion on the surrounding access roads
Out-of-town location usually means less congestion	Tends to be the same chain stores that populate shopping centres and retail parks and therefore they do not support smaller independent stores
Often room for expansion due to edge-of-city location	Land use conflict – the edge-of-city location is in high demand for other uses such as golf courses and business parks
Land values are cheaper than city-centre locations and therefore the shops are larger, holding more variety of stock	
Near suburban housing estates and therefore close to customers and workforce	

Costs and benefits of internet shopping

The area of retailing that is growing most rapidly is that of internet shopping. In 2014, almost three-quarters of all adults reported buying goods or services online, an increase of 21 per cent since 2008. The majority of this activity was carried out on desktop computers but very quickly this is moving to mobile devices such as smartphones and tablets.

Benefits	Costs
Convenient and often cheaper method of browsing and buying goods	Not everyone, particularly the elderly, has internet access
Customers can buy products not available locally	Goods may not be as expected when delivered and it may be difficult to return them
Customers can buy at any time from any location	City centre shops lose trade, which may lead to job losses and eventually closure of shops
It is less time consuming	More delivery vans increase traffic congestion and pollution
Traffic congestion in city centres is reduced	Storage of bank or credit card details online can leave customers vulnerable to fraud
Jobs are provided for those delivering products	

High street change in towns and cities

In the past ten years the UK has seen a decline of its high streets with an increasing number of vacant outlets and the disappearance of well-known retailers such as BHS. Using a variety of strategies, the high streets are changing, with the aim of tempting the shoppers back.

Example of high street change: Lancaster

After experiencing a reduction in the number of shoppers to the city centre, Lancaster's local council has made improvements to the shopping environment to attract shoppers:

- New paving in some of the pedestrianised areas giving a fresh, clean look and reducing trip hazards.
- New street furniture made from stronger materials to ensure it looks new for longer.
- New signs in pedestrianised areas to help shoppers and visitors to find their way around.
- Trees planted in the pedestrianised area to enhance the natural environment.
- New secure bike racks to encourage cyclists to visit more easily.
- A market day to attract new traders to add to the variety of independent retailers.

Now test yourself

TESTED

1. Give three economic factors that cause a decline of the traditional CBD shopping centres.
2. Describe the benefits of internet shopping.
3. Choose two strategies that are used to encourage people back to CBD shopping and explain how they attract people to shop on the high street.

Revision activity

Think about your closest urban town or city centre retail area. What changes have taken place there in the past five years? List these down the left-hand side of a piece of paper. On the right-hand side explain why each change will encourage more people to shop in that area.

Exam practice

'The characteristics of a sustainable urban community as shown by Egan's wheel are not all achievable in one urban area.' Give reasons for and against this statement. [8]

Urban issues in contrasting global cities

What are the global patterns of urbanisation?

REVISED

Patterns of urbanisation

The size of urban areas across the world is growing in terms of both physical area and number of people (**urbanisation**). Just over half of the world's population now live in urban areas, and this percentage is set to increase further. Cities in **newly industrialised countries (NICs)** are growing at a particularly fast rate. The table below shows the world's largest **mega-cities** in 2015.

City and country	Population in 2015
Tokyo, Japan	38 million
Delhi, India	26 million
Shanghai, China	24 million
São Paulo, Brazil	21 million
Mumbai, India	21 million

The distribution of the largest mega-cities is overwhelmingly in Asia. Europe does not appear even in the top ten mega-cities as they tend to be more frequent in NICs. In the future, it is expected that both China and India will continue to have a greater number of mega-cities, which is not surprising as these countries have the largest populations. Nigeria is also projected to add a further 212 million urban dwellers by 2050.

Urbanisation The growth of towns and cities

Newly industrialised countries (NICs) Middle-income countries where the pace of economic growth is faster than in other developing countries

Mega-cities Cities with over ten million residents

Exam tip

When answering a question where you need to describe the distribution in relation to data given in a table, remember to:

- give the overall pattern
- give examples of the highest and lowest
- quote some of the figures from the table.

Spotting anomalies in the data is also creditworthy.

Why global cities are important

As a result of **globalisation**, places around the world are now more connected than ever before in terms of economics, trade, society and culture, and politics. **Global cities** have become key places of connection, often with the following features:

- **finance and trade**: location of stock exchange and bank headquarters
- **governance**: location of central government, international organisations such as the United Nations and the EU, and the headquarters of multinational corporations (MNCs)
- **diversity**: global cities attract large numbers of migrants from within the country and beyond
- **media**: location of media corporations
- **cultural centres**: location of a wide range of entertainment venues
- **innovation**: location of top-rated universities and research.

Globalisation The global web of links between countries involving people, trade, ideas and cultures

Global cities Cities that play an important role in the global economic system of finance and trade

Distribution of global cities

Although global cities are distributed widely across the world this is not an even distribution. There are areas where they are clustered and other parts of the world where there are very few. For example:

- North America, Western Europe and South Asia have clusters of global cities
- Africa has very few global cities, with just six of them
- India has eight
- China has fourteen.

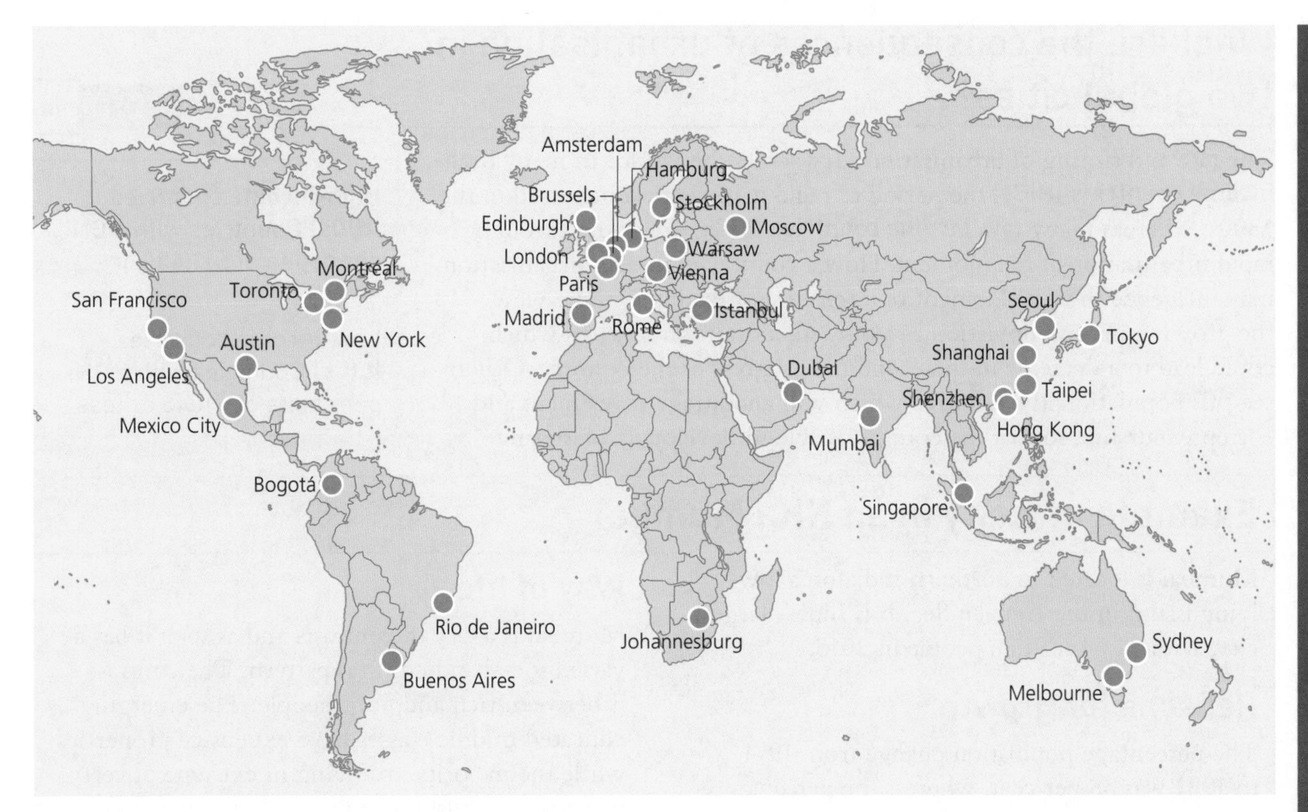

Figure 8 Global cities.

Changing pattern of global cities over time

The pattern of global cities is not static and is subject to change. To keep their global city status, each city must continue to develop and strengthen its links with other places around the world. As cities grow and develop within NICs, they are hoping to become new global cities. They look to exploit new trade routes by becoming a vital connection. If an existing global city does not continue to develop then it may well lose its global city status.

Now test yourself

TESTED ☐

1 Explain the difference between a mega-city and a global city.
2 What do you understand by the term 'globalisation' and why does this cause urbanisation to occur?
3 Why might a city lose global city status?

Revision activity

Pick three global cities and make a list of the characteristics that you think make them global cities.

Exam practice

1 Look at Figure 8, showing the locations of global cities. Describe their distribution. [4]
2 Explain why cities in NICs are growing at the fastest rate. [4]
3 Describe the characteristics that cities need to have to become classified as a global city. [4]

What are the consequences of urbanisation in two global cities?

REVISED

The rate and timing of urbanisation vary across the world. In many **high-income countries (HICs)** the period of rapid urbanisation occurred in the 1800s, whereas many **low-income countries (LICs)** are going through rapid urbanisation at the moment. How a country embraces urbanisation may influence the development of its cities. For example, some view the growing urban population as a burden, a drain on services which could lead to a decrease in the economic progress of the country. Others see this population as a resource which will encourage investment and entrepreneurship, so that the country is able to develop at a faster rate.

High-income countries (HICs) Countries with a GNI per capita of $11,456 or more

Low-income countries (LICs) Countries with a GNI per capita of $1045 or less

Example of a city in an NIC: Mumbai

Mumbai is located in northern India on a low-lying island in the Arabian Sea. It is India's largest city, with 21.04 million people in 2015.

Reasons for growth

The percentage population change from 1971 to 1981 was 38 per cent, whereas the percentage population change from 2001 to 2011 was 4.7 per cent. The reasons why Mumbai has grown to become a global city are:

- **Natural population change**: the fertility rate of women in Mumbai in 1974 was 4, which has reduced to 1.8 in 2013. Natural change would have been a contributing factor in the 1970s and 1980s, but less so now.
- **Migration**: the **pull factors** for Mumbai are cheap rail travel, jobs and better training opportunities. The **push factors** from the surrounding countryside areas are poor standard of housing, healthcare and sanitation.
- **Connections**: Mumbai is the financial capital of India as it is home to India's stock exchange. It is also home to large MNCs such as Tata Steel and home to the Bollywood film industry. It has an international airport and the major port of Nhava Sheva.
- **Historic or recent change**: the growth is largely historic. The city is still growing, but not as rapidly as it was between 1971 and 1991.

Way of life

Mumbai is a city of contrasts and as such it has a varied social and cultural pattern. The contrast is between rich and poor people. The emerging educated middle classes have expensive properties while the majority are living in extreme poverty in slums and bustees and work in the **informal economy** in such roles as street vendors and rubbish recyclers.

Current urban challenges

There are two challenges that face Mumbai:

- **Reducing poverty and deprivation**: with such a large proportion of people living in slums, Mumbai has millions of people who are stuck in a cycle of deprivation.
- **Housing**: the majority of people in Mumbai live in slums, are pavement dwellers or live in chawls (old four- or five-storey tenement buildings with very basic shared facilities). These dwellings suffer from overcrowding, and are at risk of collapse, flooding or fire. Two possible solutions are:
  - Self-help projects which involve giving the residents help to improve their own living conditions, for example connecting to the mains water supply.
  - Wholesale clearance by demolishing the existing dwellings and building purpose-built high-rise buildings.

Pull factors Factors that attract people to a place

Push factors Factors that make someone want to leave a place

Informal economy Forms of employment that are not officially recognised, for example the money earned from irregular jobs or from working for yourself on the streets

Example of a city in an HIC: Cardiff

- Cardiff is located in south-east Wales and is Wales' largest city and its capital.
- It has a population of 346,000 and a large sphere of influence, with 1.49 million people living within 32 km of its city centre.
- It is very well connected, with a mainline railway and bus station in addition to the M4 motorway running north of the city.

Reasons for growth

Cardiff has been the fastest growing core economic city (outside London) in the UK over the past decade. Cardiff is projected to grow by a further 26 per cent (or 91,500 people) over the next twenty years. The reasons for this are:

- **Natural population change**: the fertility rate was falling in the UK until 2001–02. However, it has been rising since then and this, coupled with the increase in life expectancy, leads to an increase in natural population change.
- **Migration**: the pull factors attracting people to Cardiff are the availability of jobs, good education and research facilities, and a thriving tourist industry. The push factors from the surrounding region are mainly due to lack of jobs in the south Wales valleys.
- **Connections**: being the capital city of Wales, Cardiff has well-developed connections to the local region and to the rest of the UK and Europe. Its port was the main point of exporting south Wales' coal early in the twentieth century. It has an international airport and the M4 links it to London in just over two hours.
- **Historic or recent change**: the rate of growth of Cardiff has increased in recent years, with an increase of 40,000 people between 2001 and 2011.

Way of life

Cardiff is a multicultural city and as such people across the city can live very different lives. Two of the causes of this are:

- **Ethnic minorities**: economic migrants have been coming to Cardiff since the 1800s. This has led to eight per cent of Cardiff's present population being from ethnic minorities, giving a great mix of food, culture, religion and language.
- **Income levels**: the range of incomes that people earn when working in Cardiff is vast, from the minimum wage up to six-figure salaries.

Current urban challenges

The main urban challenges for Cardiff are:

- reducing poverty and deprivation, for example in Butetown
- reducing traffic congestion, for example the A470 road
- regeneration of the CBD; shop closures on Queen Street.

Now test yourself

TESTED

1. Describe the locations of Mumbai and Cardiff.
2. Give four factors as to why Cardiff and Mumbai have grown into global cities.
3. Think of as many differences as you can to show how life is different for the rich and poor of Cardiff. Repeat this for Mumbai.
4. What are the similarities and differences between the consequences of urbanisation in Cardiff and Mumbai?

Revision activity

Using an A5 card, create a fact file for Cardiff so that you can learn the specific facts about that city and the issues affecting it. Repeat this for Mumbai.

Exam practice

1. What is meant by the term 'global city'? [2]
2. Why are challenges in informal settlements difficult to overcome? [6]

Exam tip

If an exam question asks you to compare, try to use connective words that link the two aspects in the same sentence, for example, whereas, compared to, however and so on.

How are global cities connected?

REVISED

The term **globalisation** was first used in the 1950s to refer to the telephone making global communications much easier. Today globalisation is much more than global communication and relies on global cities to make connections for trade, movement of people, media and so on. It is only through multiple types of connections that goods can be sold internationally, cultures can be shared and people can migrate. There are three main types of connection: through transport, trade and tourism, and media and communications.

Globalisation Flows of people, ideas, money and goods making a global web that links people and places

Infrastructure The basic structures and services needed by any society, for example roads, railways, water and electricity supply

Connections through transport

The world's major logistical companies have developed an elaborate and flexible transport system which includes airports, ports, airlines, shipping and rail to move goods and people to any part of the globe. As the number of routes has increased, a more sophisticated method of connecting places has emerged through using transport hubs. These enable greater flexibility within the transport system, through a concentration of flows. For instance, a point-to-point network involves sixteen independent connections, each to be serviced by vehicles and **infrastructure**. By using a hub-and-spoke structure, only eight connections are needed.

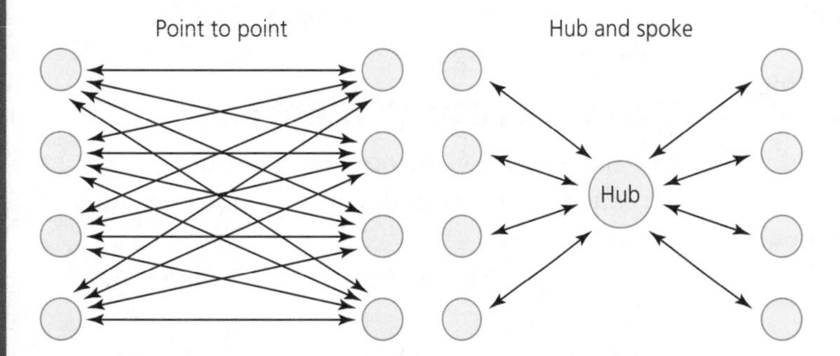

Figure 9 A point-to-point network and a hub-and-spoke structure.

Connections through trade and tourism

Locations around the world have been connected by trade for hundreds of years. With the growth of trade and the development of MNCs these connections have become much stronger. For example Tata Steel, an Indian MNC with its headquarters in Mumbai, has manufacturing plants all over the world including the UK.

Tourism has also led to countries becoming more connected. Long-haul flights to remote destinations are far more common these days. Tourists can even travel to Antarctica for a holiday. The number of short-haul flights has also increased in terms of the frequency of flights and number of destinations reached. This has been greatly aided by the increase in budget airlines which fly to smaller regional airports as well as the central hubs in many countries. This movement of people leads to a rich exchange of culture and understanding of other people's way of life, which enables us to feel more connected to each other.

Connections through media and communications

The media has seen the fastest change of connections in the twenty-first century. Social media such as Facebook, Twitter and Instagram allow anyone with an internet connection to broadcast to the world what is happening to them 24 hours a day. As such, people across the world can connect at the touch of a button and exchange information. This increases awareness of events around the world and also different cultures. This has been enabled largely through the development of the internet and high-speed broadband connections so that people can communicate in real time.

How Mumbai and Cardiff are connected to the world

Infrastructure	Mumbai's connections	Cardiff's connections
Airport	Chhatrapati Shivaji International Airport: carrying over 36 million passengers to 45 different countries in 2015	Cardiff International Airport: with over 25,000 flights in 2015 going to 60 different countries
Port	Mumbai Harbour: handles 57 million tonnes of cargo annually	Port: handles 2.5 million tonnes of cargo annually
Internet	Broadband coverage of varying speeds	Superfast fibre-optic broadband widely available
Media	Bollywood film industry is based in Mumbai and its movies are seen all over the world	BBC Cymru Wales is based in Cardiff making TV programmes broadcast to the whole of the UK

Now test yourself

1. What is a transport hub?
2. Why are transport hubs a more efficient way of moving goods and people around the world?
3. For two cities you have studied, make a list of the key transport connections that they have.

Revision activity

Draw a double-bubble diagram like the one on page 32 to show how people's lives in Cardiff and Mumbai benefit from global connections through: a) trade and tourism and b) media and communications.

Exam practice

1. Describe what a transport hub is. [2]
2. Explain why tourism enables global cities to be connected. [4]
3. For two global cities that you have studied, describe the connections that they have to the rest of the world and decide which type of connection is most important to that city. [8]

Theme 3 Tectonic Landscapes and Hazards

Tectonic processes and landforms

How do tectonic processes work together to create landform features?

REVISED

Plate movement and boundaries

The outer layer of the Earth is called the crust. The crust may be divided into two types:

- continental crust, which on average is 35 km thick
- oceanic crust, which is much thinner at between 6 and 8 km thick.

The crust and the uppermost mantle, which make up the hard and rigid outer layer of the Earth, is known as the lithosphere. The lithosphere is split into **tectonic plates**.

Tectonic plates move relative to one another. High temperatures in the Earth's core cause convection currents and a plume of hot **magma** rises through the mantle. The oceanic plate is forced upwards by the rising magma and torn apart to produce a mid-ocean ridge. The semi-molten rock then spreads out, carrying the plate above with it. Oceanic crust converges with continental crust and flexes downwards under the continental crust, forming an ocean trench. Magma then cools and sinks back into the mantle.

Tectonic plates The Earth's crust and upper part of the mantle is split up into large sections

Magma Molten rock located below the Earth's surface within the mantle or crust

Plate boundary or margin The place where two or more plates in the Earth's crust meet

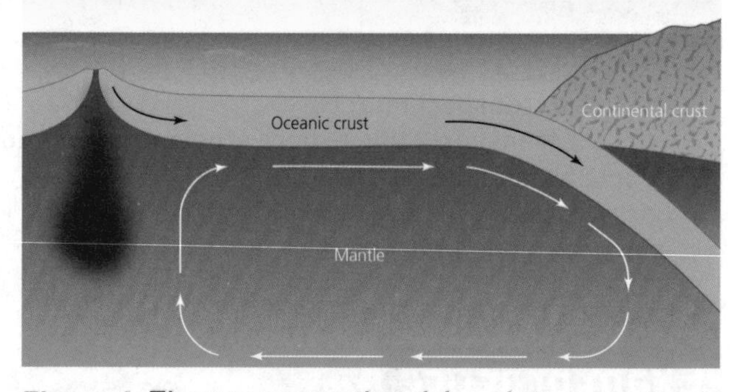

Figure 1 The processes that drive plate movement.

Global distribution of tectonic activity

Plate movement causes earthquakes and volcanoes. The point where two plates meet is called a **plate boundary or margin**. The map in Figure 2 shows the world's tectonic plates and plate margins.

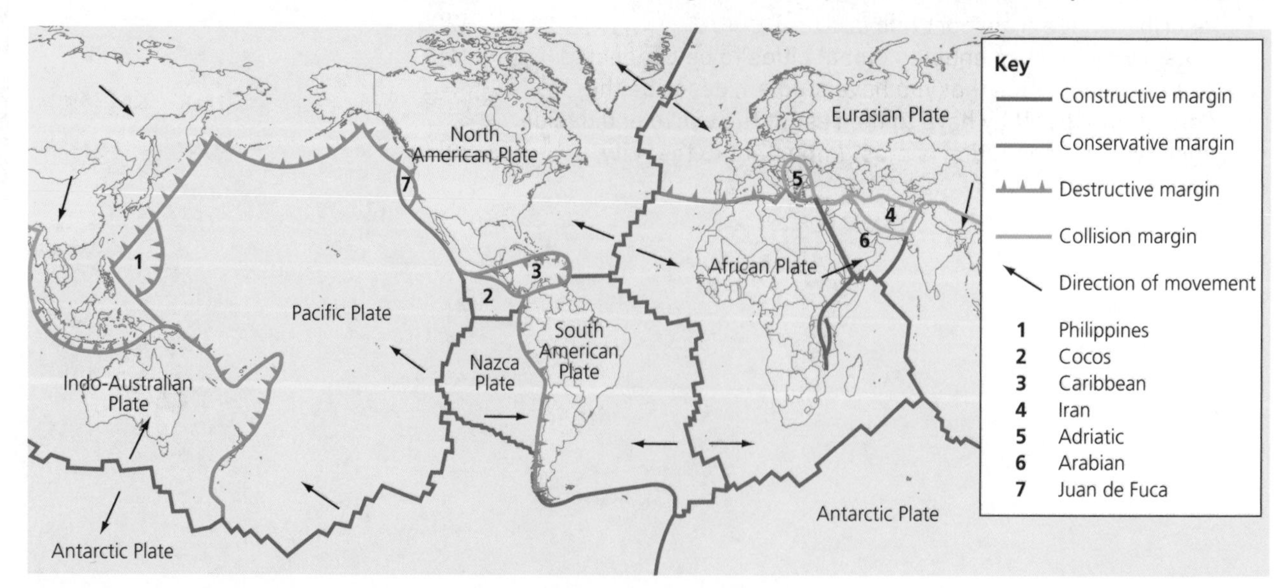

Figure 2 Plate margins and the direction of plate movement.

Large-scale tectonic processes at the plate margins

Earthquakes and **volcanoes** are most likely to occur near plate margins:

- Where plates converge a **destructive margin** is formed.
- Where plates diverge a **constructive margin** is formed.

Convection, subduction, convergence and divergence

The processes that cause plate movement are:

- Radioactive decay in the Earth's core heats the magma in the mantle above and creates **convection** currents like water boiling in a saucepan. The convection currents move the plates. Where convection currents **diverge** near the Earth's crust plates move apart. Where convection currents **converge** plates move towards each other.
- When an oceanic plate and continental plate collide the denser oceanic crust is forced underneath the continental crust. The process where the oceanic plate pushes under the continental crust and slides back into the mantle is known as **subduction**.

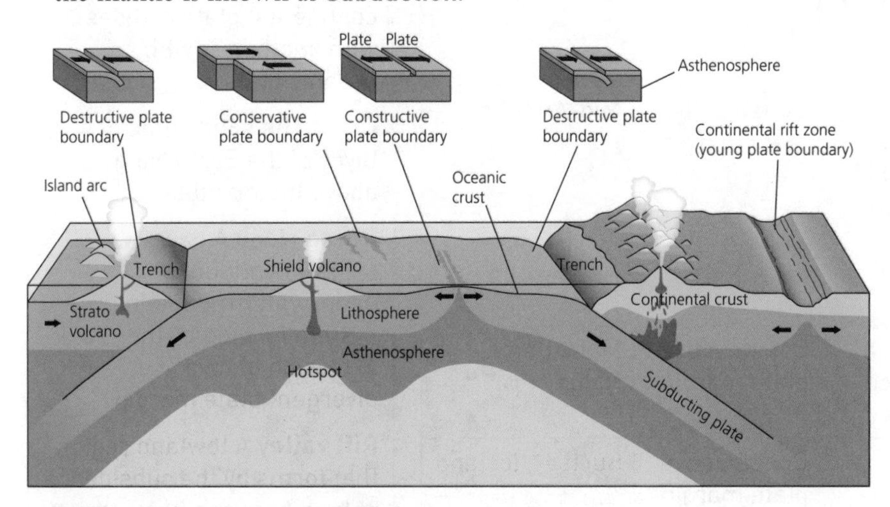

Figure 3 Constructive and destructive margins

Earthquake A tremor of the surface of the Earth resulting from shockwaves generated by the movement of rock masses within the Earth, particularly near boundaries of tectonic plates

Volcano A mountain created by the eruption and deposition of lava and ash from a vent in the ground

Destructive margin A plate boundary, sometimes called a convergent or tensional plate margin, where oceanic and continental plates move towards each other

Constructive margin A plate boundary, sometimes called a divergent plate margin, where the crustal plates move apart from each other

Convection When heat in a gas or liquid is transferred from a warmer to a cooler place by upward movement

Subduction The process in which an oceanic plate collides with and is forced down under another crustal plate and drawn back into the mantle

Now test yourself

TESTED

1. Explain why the knowledge of the asthenosphere is vital to our understanding of the theory of plate movement.
2. What is the difference between a constructive and destructive plate margin?
3. Outline the processes that take place at a destructive plate margin where an oceanic plate meets a continental plate.
4. 'Ocean crust is created and destroyed. Continental crust is folded, crushed and compressed, but not destroyed.' Use evidence from Figure 3 to explain this statement.

Exam tip

It is important to focus on the command word when answering a question. The two most commonly used command words are describe and explain. Describe invites you to 'paint a picture' of what something is like by using lots of adjectives. Explain directs you to say why it is like that.

Revision activity

1. On a sheet of A5 card make a copy of Figure 2. Annotate your diagram to explain why plates move over the surface of the Earth.
2. Draw a table and list the names of the seven large plates.
3. Describe the location of the constructive and destructive plate boundaries.

What is happening at the plate margins?

REVISED

Large-scale features formed by tectonic processes

Tectonic processes result in the formation of distinct landscapes. Examples of the features formed by these processes include:

Large-scale feature	How formed	Location	Example
Ocean trench	Where subduction takes place	Destructive plate margin	Mariana Trench, Western Pacific Ocean
Fold mountains	The continental crust is crushed and folded upwards	Destructive plate margin	Andes Mountains, South America
Explosive volcanoes	As the oceanic plate sinks, it melts and the molten magma finds its way to the surface	Destructive plate margin	Mount Merapi, Indonesia
New **crust**	Where two oceanic plates move apart, the space between the diverging plates is filled with magma	Constructive plate margin	Mid-Atlantic
Ocean ridge	As lava cools, a ridge is formed under the sea	Constructive plate margin	Mid-Atlantic Ridge
Submarine volcanoes and volcanic islands	Submarine volcanoes sometimes rise above the surface of the sea to create volcanic islands	Constructive plate margin	Surtsey, Iceland
Rift valley	Where two continental plates pull apart	Constructive plate margin	Thingvellir, Iceland

Ocean trench A long, narrow, deep depression in the ocean floor formed at a subduction zone where the denser plate is forced below the less dense one

Fold mountains Mountains that form by folding of layers in the upper part of the Earth's crust. Most commonly formed where a continental plate collides with another or with an oceanic plate

Crust The solid outermost layer of the Earth lying above the mantle

Ocean ridge A narrow, largely continuous underwater mountain system formed by the extrusion of lava at a divergent plate margin

Rift valley A lowland region that forms by the subsidence of land between two parallel faults where the Earth's tectonic plates move apart or rift

Example of an ocean trench: Mariana Trench

The Mariana Trench is located at a destructive plate boundary where the oceanic Pacific Plate subducts under the oceanic Philippine Plate. The trench is the deepest spot in the world's oceans and includes the Challenger Deep which reaches a depth of 10,994 m.

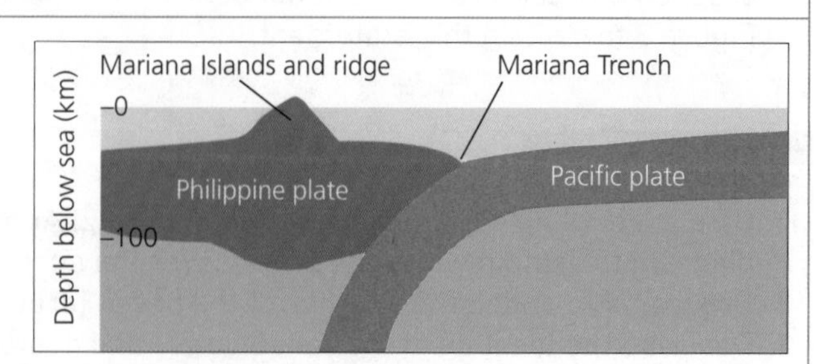

Figure 4 The Mariana Trench.

Example of a rift valley: Thingvellir, Iceland

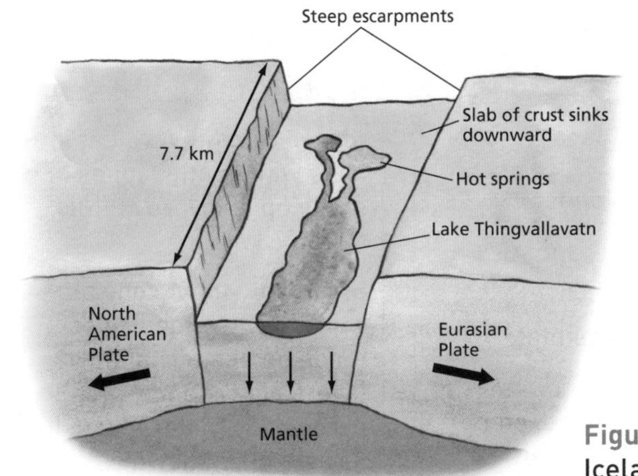

The Mid-Atlantic Ridge is located along the floor of the Atlantic Ocean. Iceland is one of the few places where it juts out of the surface of the ocean. Iceland is being gradually torn in two as the North American and Eurasian Plates diverge. This is best seen in Thingvellir National Park:

- a rift valley 7.7 km long has been formed
- the valley walls are pulling apart at an average of 7 mm a year
- the valley floor is subsiding by 1 mm a year.

Figure 5 **Formation of the rift valley at Thingvellir, Iceland.**

Now test yourself

TESTED

1. Give the names of two landforms found at a destructive plate margin and two landforms found at a constructive plate margin.
2. Explain why the type of volcano found at a destructive margin is different from one found at a constructive plate margin.
3. Explain why rocks get increasingly older the further away from the ocean ridge you travel.

Revision activity

Sketch a diagram of a destructive plate margin. Label the following features:

- fold mountains
- subduction zone
- ocean trench
- magma
- explosive volcano
- ocean crust
- continental crust.

Volcanic hotspots

A hotspot is a small area of the Earth's crust which has an unusually high amount of volcanic activity. Iceland has formed above a hotspot on the Mid-Atlantic ridge, although most hotspots are located away from plate boundaries, for example the Hawaiian Islands. One suggestion about the way in which hotspots form is:

- Intensive radioactivity in the Earth's interior creates a huge column of upwelling magma, known as a mantle plume.
- The plume pushes upwards, melting and pushing through the crust above.
- The plume lies at a fixed position under the tectonic plate. As the plate moves over this 'hotspot', the upwelling magma creates a succession of new volcanoes that migrate along with the plate.

Example of a hotspot: the Hawaiian island chain

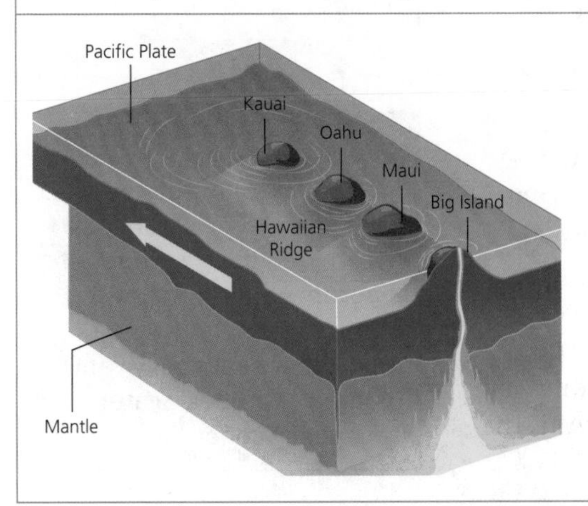

The Hawaiian Islands are the tops of volcanic mountains formed by eruptions of fluid lava over millions of years. Some of these mountains are more than 9000 m above the seafloor. The islands represent the visible part of an ocean ridge, the Hawaiian Ridge. Over 70 million years the movement of the Pacific Plate over a stationary hotspot has left a chain of volcanoes and the creation of the Hawaiian Islands.

Figure 6 **Hotspot formation of volcanic island chains.**

Distinctive features of a volcanic landscape

Volcanoes occur where weaknesses in the Earth's crust allow magma, gas and water to erupt on to the land and sea bed.

Larger-scale features

These include **shield volcanoes**, **stratovolcanoes** and **caldera**.

Feature	Shield volcano	Stratovolcano
Example	Mauna Loa, Hawaii	Mount Merapi, Indonesia
Location	Constructive plate margins and above hotspots	Destructive plate margins
Shape	Circular shape, gently sloping sides	Conical shape, steep-sided
Formation	Basaltic magma, high in temperature, with a low silica and gas content finds its way to the Earth's surface through cracks in the crust. Magma produces fluid lava which flows long distances before solidifying. Frequent eruptions, gentle oozings of lava form large cone-shaped mountains	Lava is acid, has high viscosity and cools quickly. Explosive eruptions of ash, lava and lava bombs form a cone-shaped volcano with steep sides as the lava doesn't flow very far before it solidifies
Composition	Lava with no layers	Alternate layers (strata) of ash and lava

A caldera (from the Spanish word for cauldron) is a large-scale volcanic crater that could be several kilometres in diameter. It is formed either:

- when a magma chamber is emptied and the roof collapses
- through a massive explosive volcanic eruption.

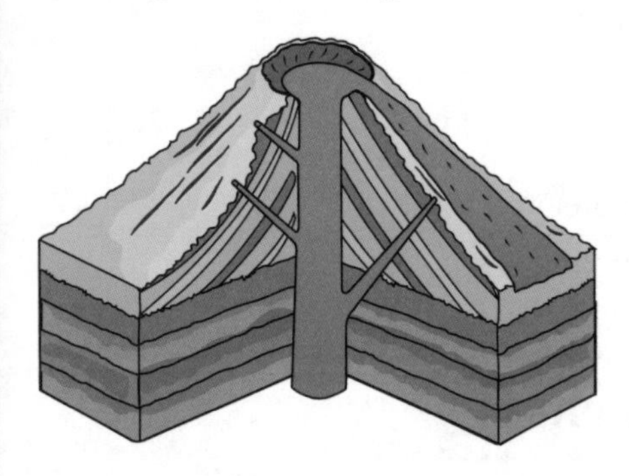

Figure 7 Stratovolcano.

Shield volcano A volcanic cone with gentle slopes made up of layers of fluid basaltic lava

Stratovolcano A conical volcano with steep sides made up of alternating layers of lava and pyroclastic material, such as ash. Also known as a composite volcano

Caldera A volcanic cone where the original top and centre have been removed, either through a massive eruption or through collapse, leaving the base of the cone as a large ring-shaped ridge

Cinder cone A steep-sided conical hill formed by the explosive eruption of cinders (glassy volcanic fragments) that accumulate around a vent

Lava tube A hollow tunnel formed when the outside of a lava flow cools and solidifies and the molten material passing through it is drained away

Geyser A hot spring that is under pressure and erupts, sending jets of water and steam into the air. The heat results from the contact of groundwater with magma bodies

Smaller-scale features

Smaller-scale features include:

- Hot cinders released when lava erupts cool quickly and build up around the vent forming a steep, round hill known as a **cinder cone**.
- A **lava tube** forms beneath the surface of the ground when low-viscosity lava develops a hard crust through which lava flows.
- A **geyser** is a vent in the Earth's surface that periodically ejects a column of hot water and steam. A geyser erupts when superheated groundwater, confined at depth, becomes hot enough to blast its way to the surface. Old Faithful, located in Yellowstone Park, USA, is the world's best-known geyser.

Revision activity

Make a copy on a sheet of A5 card of a stratovolcano, as shown in Figure 7. Label this diagram with the following features: ash, lava flow, main vent, secondary vent, magma chamber, crater, secondary cone.

Now test yourself

TESTED

1 Complete the following table giving two ways in which a stratovolcano is similar to a shield volcano and two ways in which it is different.

Similar	Different

2 Name and explain the formation of one large-scale feature of a volcanic landscape and one small-scale feature.
3 Explain why secondary impacts may last for several years after the event.

Exam practice

Describe how tectonic processes at a destructive plate margin have resulted in the formation of any large-scale feature, such as an ocean trench or a volcano. [4]

Exam tip

Diagrams need to be clear and detailed and must highlight important features. Annotation is more than labelling, it is a command word that demands that you add explanatory notes to your diagram.

Vulnerability and hazard reduction

What are the impacts of tectonic processes?

REVISED

Tectonic processes may cause earthquakes, tsunamis or volcanic eruptions. The immediate consequences of an event are its primary impacts, such as people inside a collapsing building during an earthquake. Secondary impacts result from the primary event, for example destroyed buildings lead to people becoming homeless. Secondary impacts may last for many years after the event.

Magnitude of volcanic eruptions and earthquakes

Measuring the strengths of volcanic eruptions is challenging since they produce different materials and have different durations. In 1982, the **volcanic explosivity index (VEI)** was devised. The index measures the volume of pyroclastic material ejected by the volcano, the height of the eruption column and the duration of the eruption.

The strength, or **magnitude**, of an earthquake was traditionally measured using the **Richter scale**. The largest earthquake ever recorded was 9.5 on the Richter scale in 1960 in Chile. The moment magnitude scale (MMS or M_W) was introduced in 1979 as a successor to the Richter scale. It measures the distance moved by a fault and multiplies it by the force needed to move it.

Each of these measures uses a logarithmic scale, with each number on the scale ten times the magnitude of the one before it.

Volcanic explosivity index (VEI) A measure of the explosiveness of volcanic eruptions. It measures how much volcanic material is ejected, the height of the material thrown into the atmosphere and how long the eruptions last. The scale is logarithmic on a scale of 1–8

Magnitude A quantitative measure of the size of an earthquake using the Richter scale

Richter scale A measure of the magnitude of an earthquake. It uses a logarithmic scale, that is each level is ten times stronger than the one below, from 1 to 10

Physical factors that increase vulnerability to volcanic hazards

Volcanic activity results in a number of localised and large-scale hazards.

Hazard	Characteristics	Scale
Lava flows	Molten rock flows down the sides of a volcano. Lava from shield volcanoes flows more quickly and travels further	Local – may travel several kilometres and threaten towns and villages in their path
Lahars	Volcanic mudflows consisting of a mixture of ash and water, from rain, melted snow and ice, which travel at great speed down the mountain	Local – may travel several kilometres and threaten towns and villages in their path
Ash clouds	Ash thrown high into the atmosphere shuts out the sun and when it settles covers the ground, buildings, crops and power lines in a layer of ash which could be over a metre thick	Large – may reach 10–15 km high and spread over thousands of kilometres
Pyroclastic flows	Burning clouds of gas and ash, with temperatures up to 1000 °C, travelling down the mountain at speeds of up to 200 km per hour	Local – may travel several kilometres

Lava flow A stream of lava flowing from a volcanic vent

Lahar Mudflow associated with volcanic activity. Surface water mixes with volcanic ash to produce the lahar

Ash cloud A large cloud of smoke and debris that forms over a volcano after it erupts

Pyroclastic flow The cloud of gas, ash, dust, stones and rocks emitted during a highly explosive volcanic eruption

Revision activity

Research the news reports on the Merapi eruption. Imagine you are a reporter covering the after-effects of the eruption, and write a script for a 60-second TV news report on how the eruption is affecting a family that once lived close to the volcano.

Example of a volcanic eruption: Mount Merapi, Indonesia

Mount Merapi is a stratovolcano located in Java. Merapi erupted in 2010 with a magnitude of 4 on the VEI. The region is densely populated.

Impacts	Social	Economic and environmental
Primary	• Ash clouds caused breathing problems • 353 deaths, largely from pyroclastic flows • 570 people injured • 320,000 people evacuated from the area	• Ash falls destroyed crops and 1900 farm animals died • Ash clouds caused hundreds of flights to and from Indonesia to be cancelled • 27 million m^3 of ash and rock were deposited in the River Gendol
Secondary	• An area 20 km around the volcano was declared a danger zone • Thousands of people spent weeks living in 700 emergency shelters • There were not enough toilets or clean drinking water	• 1300 ha of farmland were abandoned • Food prices rose. The people of Indonesia are poor and could not afford the higher prices • $700 million revenue was lost due to agricultural losses and fewer tourists

Now test yourself

TESTED

1. Describe two tectonic hazards faced by people who live close to an active volcano.
2. Why is Merapi described as a dangerous volcano?

Physical factors that increase vulnerability to earthquakes

Most earthquakes are associated with movements along a plate margin. Earthquake hazards include:

- Ground movement and shaking cause bridges and buildings to collapse and underground pipelines to rupture.
- Soil **liquefaction** damage buildings' foundations and results in them sinking.
- Landslides bury people, livestock and buildings.
- **Tsunamis** cause severe damage and death in low-lying coastal areas.

How **vulnerable** to tectonic activity a region is depends on:

- Magnitude: the stronger the hazard the more severe the impacts.
- Duration: the longer a hazard lasts the more severe the impacts are likely to be.
- Predictability: hazards that hit with no warning are going to have more serious results.
- Regularity: if hazards happen often and in quick succession, for example an earthquake followed by multiple **aftershocks**, then the severity is likely to be greater. Communities do not have the **capacity** to recover before the next earthquake hits.

Example of an earthquake: Italy

An earthquake of 6.2 magnitude on the Richter scale struck in central Italy at 03.36 (local time) on 24 August 2016. The earthquake was caused by a continental collision between the African and Eurasian tectonic plates. The **focus** was at a shallow depth of approximately 6 km.

Impacts	Social	Economic and environmental
Primary	295 deaths and 400 injuries The earthquake hit during the summer holiday season, the population of the region was much higher than at other times of the year and the death toll included tourists	The cost of repairs, including rebuilding costs, is estimated at around $11 billion A 4.8 magnitude earthquake followed at 06.28 (local time) on 26 August, causing more damage to collapsed buildings and hampering rescue efforts
Secondary	Over 500 aftershocks left over 2500 people homeless In Amatrice, a village near the **epicentre**, over half the buildings were destroyed, including the historic town	The tourist industry is likely to take many years to recover Following the earthquake, the Italian press has criticised the government over building regulations, since historic towns do not have to conform to anti-quake building laws

Liquefaction Occurs when vibrations cause soil particles to lose contact with one another. As a result soil behaves like a liquid, has an inability to support weight and can flow down very gentle slopes

Tsunami Also known as a 'seismic sea wave'. When an earthquake lifts or drops part of the ocean floor the water above rises and forms a series of waves called a tsunami. In the open ocean, the tsunami wave is only about a metre high. As it approaches shallower water, near to the coast, the wave builds to a great height

Vulnerability The potential to be harmed by a natural hazard. Some people and places are more vulnerable than others

Aftershocks Ground tremors occurring after a major earthquake but associated with the same focus

Focus The source of a shockwave, which can be at varying depths

Capacity The ability of a country or region to react to and recover from a natural hazard

Epicentre The point on the Earth's surface immediately above the focus

Social and economic (human) factors that increase vulnerability

- Wealth: poor people are less able to afford housing that can withstand extreme events and are less likely to have money or insurance policies that can help recovery.
- Education: when populations are literate, written messages can be used to spread information either before the event or to issue warnings and give advice during an event.
- Governments: can support education and awareness, and can pass building regulations.
- Age: children and elderly people are more vulnerable. They have fewer financial resources and are frequently dependent on others for survival.
- Health: a healthy person is more able to escape the dangers and recover after the event.
- Population density: the greater the number of people who live in the area, the more severe the impact.
- Time of day or day of the week: influences whether people are at home, at work or travelling. An earthquake at rush hour in a densely populated urban area could have devastating effects.
- Emergency services: richer countries usually have well-trained and well-resourced response teams that can rescue and treat people following a disaster.

Revision activity

On a sheet of A5 card, build up a Five Ws case study of the Italy earthquake.

The Five Ws are questions which will help you to gather information or problem-solve. They will provide you with a complete story of the earthquake.

- **What** happened?
- **When** did it take place?
- **Where** did it take place?
- **Why** did it happen?
- **Who** did it affect?

Each question should have a factual answer.

Now test yourself

TESTED

1. Explain why the shallow depth of the Italian earthquake was important to the amount of damage that was caused.
2. Suggest why the time of day (and time of year) an earthquake strikes is important.
3. Describe the primary and possible secondary impacts of the Italian earthquake.
4. Italy is considered a wealthy country. If the same event occurred in a less developed country how would the short-term and long-term impacts be different?

Figure 8 Impact of the tsunami on the coast of Japan.

Example of an earthquake and tsunami: Tōhoku, Japan

On 11 March 2011, a magnitude 9 earthquake occurred off the coast of Tōhoku, Japan. Less than half an hour later the coast was hit by a tsunami. Waves reached 10 m high and travelled 10 km inland. Over 20,000 people died. The earthquake happened because of a build-up in strain energy as the Pacific Plate subducted under the Eurasian Plate.

Revision activity

Complete a diagram, such as the one below, summarising the physical and human (social and economic) factors that affect vulnerability. Use different colours for physical and human.

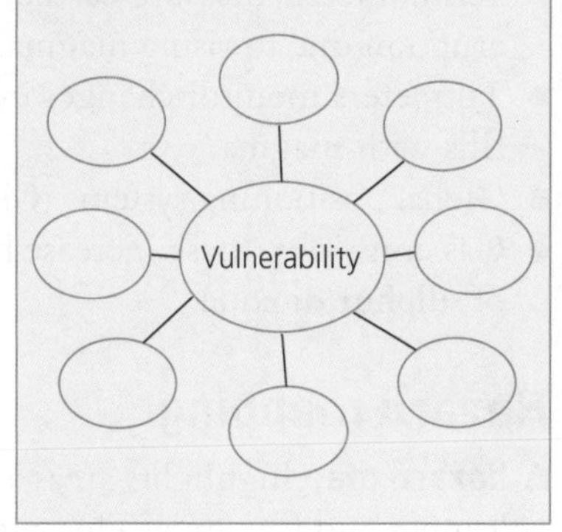

Now test yourself

TESTED

1. What is a tsunami?
2. 'People who live in low-income countries (LICs) are likely to be more vulnerable to tectonic hazards than people who live in high-income countries (HICs).' To what extent do you agree with this statement?

Exam practice

Study the photograph which shows the impact of the tsunami on the coast of Japan.

Use evidence in the photograph only to describe two ways in which the tsunami affected the lives of people who live in the coastal areas of Japan. [4]

Exam tip

You must only use evidence in the photograph. Make sure you develop points so that you gain all of the marks available in a question.

How might the risks associated with tectonic hazards be reduced?

REVISED

Prevention is not an option. This leaves two possible ways of managing tectonic hazards: **prediction** through monitoring and hazard mapping, and **preparation** including new building technology and emergency planning.

Seismometer (or **seismograph**) An instrument used to detect and record earthquakes

Hazard map A map that highlights areas affected by or vulnerable to a particular hazard

Risk The probability of a hazard event causing harmful consequences (death, injury, loss of property, damage to environment and so on)

Monitoring earthquakes, tsunamis and volcanic eruptions

Earthquakes are difficult to predict. Monitoring techniques include:

- Laser beams are used to detect plate movement.
- A **seismometer** (or **seismograph**) is used to pick up the vibrations in the Earth's crust. An increase in vibrations may indicate a possible earthquake.
- Levels of radon gas escaping from cracks in the Earth's crust can be monitored. An increase may suggest an earthquake.

Tsunami warning system:

- Following the 1960 Chilean earthquake and tsunami, the nations of the Pacific decided to set up the Pacific Tsunami Warning System (PTWS).
- The PTWS uses a network of seismometers and ocean buoys to detect earthquakes that might cause tsunamis.
- Warnings are given to local centres around the Pacific region, which warn local people via TV, radio, text messages and sirens, giving time to evacuate the area.
- In the aftermath of the 2004 Indian Ocean tsunami, which killed over 230,000 people, the PTWS has taken on additional areas of responsibility which include the Indian Ocean.

Monitoring techniques used to predict volcanic eruptions include:

- Remote sensing: satellites monitor gas emissions and use thermal imaging to study changes in the temperature of the volcano.
- Visual signs: cameras are used to look for visual signs of change in the volcano.
- Seismometers measure earthquake activity. Activity increases before an eruption due to rising magma.
- Tiltmeters monitor changes in the shape of a volcano that occur as it fills with magma.
- Global positioning systems (GPS) detect movements of as little as 1 mm.
- Gas emissions: these increase before an eruption, particularly emissions of sulphur dioxide.

Revision activity

1. Choose five ways in which volcanoes are monitored and five ways in which people can prepare for a tectonic event. Draw and label a picture of each on a sticky note. Stick them where you can see them until you can remember each one.
2. Complete a wheel of knowledge for the topics in this theme (see www.hoddereducation.co.uk/myrevisionnotes).

Hazard mapping

A **hazard map** highlights areas affected by, or vulnerable to earthquakes, volcanoes and tsunamis. This allows local authorities to:

- limit access to hazardous areas
- control development in areas at **risk** from tectonic events.

Exam tip

Questions which require extended writing are marked using a levels mark scheme. To achieve the top level and gain full marks your answer needs to demonstrate exceptional application of knowledge and understanding, a comprehensive chain of reasoning and a balanced appraisal.

Now test yourself

TESTED

1. What is a hazard map?
2. Describe two features of an earthquake-resistant building.
3. Explain why these features could reduce the risks associated with earthquakes.

Example of a hazard map: Soufrière Hills Volcano, Montserrat

This hazard map divides Montserrat into six zones. Access to areas is restricted depending on how active the volcano is:

- Zone V is out of bounds to all except scientists.
- The central zones (A–C and E): residents are on a heightened state of alert, have a rapid means of exit and have hard hats and dust masks.
- The northern zone: an area with lower risk, suitable for residential and commercial development.
- The maritime exclusion zones (E and W) exist because pyroclastic flows can travel out to sea.

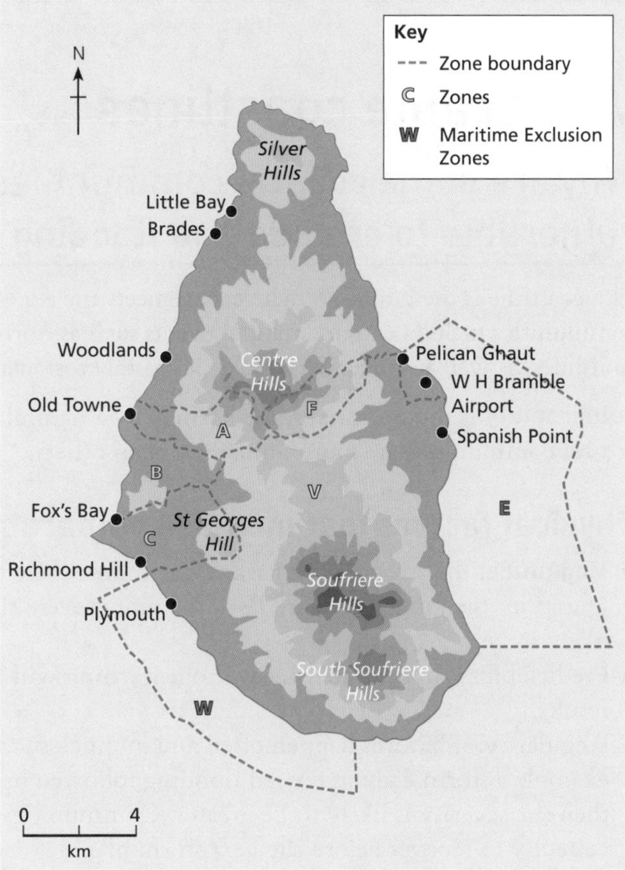

Figure 9 **Hazard map for Montserrat.**

New building technology

Earthquake-proof buildings have been constructed in many major cities, designed to absorb the energy of an earthquake and to withstand the movement of the Earth.

Emergency planning

- An exclusion zone can be set up near a volcano.
- Evacuation routes enable residents to leave the area.
- Lava flows can be diverted.
- Emergency services need to be trained and have the necessary equipment.
- People can be educated through TV and social media on what they should do to protect themselves in the event of a volcanic eruption, earthquake or tsunami. Rehearse earthquake drills, for example.
- People may put together emergency kits containing first-aid items, blankets and tinned food, and store them in their homes.
- Roads and bridges can be designed to withstand the power of earthquakes.

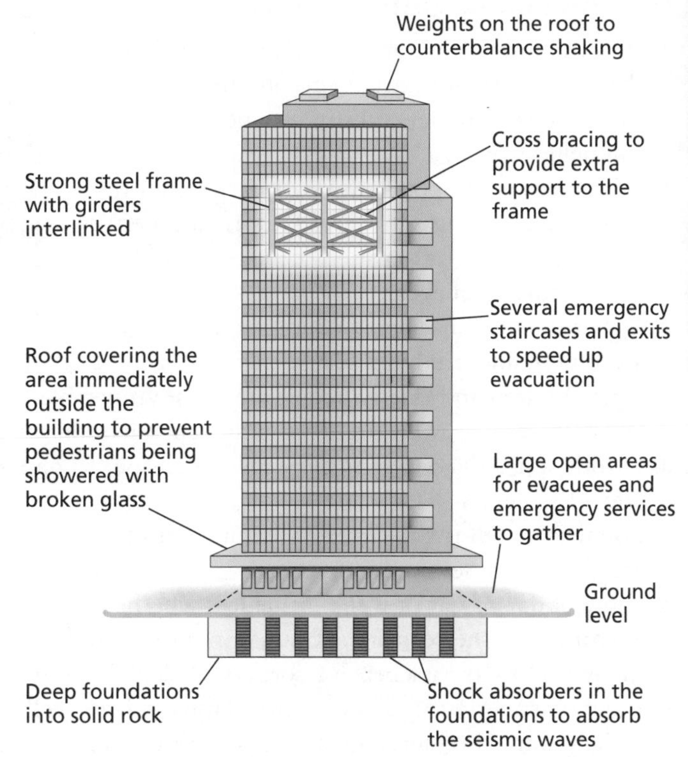

Figure 10 **Features of an earthquake-resistant building.**

Exam practice

Explain why people who live in lower income countries are likely to be more vulnerable to the impact of tectonic hazards. [6]

Theme 4 Coastal Hazards and their Management

Vulnerable coastlines

Why are some coastal communities vulnerable to erosion and flooding?

REVISED

The coastline is the boundary where land meets the sea. Threats to coastal communities include extreme natural events such as storms, tsunamis and landslides, as well as longer-term risks of coastal erosion and sea level rise.

Vulnerability is the potential to be harmed by a natural hazard. Some coastal communities are more vulnerable than others.

Vulnerability The potential to be harmed by a natural hazard. Some people and places are more vulnerable than others

Magnitude A quantitative measure of the size of a natural event such as a tropical storm

Capacity The ability of a country or region to react to and recover from a natural hazard

Physical factors that increase vulnerability

- **Magnitude**: the stronger the hazard, the more severe the impacts.
- Duration: the longer a hazard lasts, the more severe the impacts are likely to be.
- Predictability: hazards that hit without warning will have more serious results.
- Regularity: if hazards happen often and in quick succession, for example a storm causing coastal flooding followed by further storms, then the severity is likely to be greater. Communities do not have the **capacity** to recover before the next storm hits.

Social and economic (human) factors that increase vulnerability

- Wealth: poor people are less able to afford housing that can withstand extreme events and are less likely to have money or insurance policies that can help recovery.
- Education: when populations are literate, written messages can be used to spread information either before the event or to issue warnings and give advice during an event.
- Governments can support education and awareness, and can build sea defences.
- Age: children and elderly people are more vulnerable. They have fewer financial resources and are frequently dependent on others for survival.
- Health: a healthy person is more able to escape the dangers and recover after the event.
- Population density: the greater the number of people who live in the area, the more severe the impact.
- Time of day or day of week influences whether people are at home, at work or at the coastline. A tsunami in the middle of the day when many visitors are at a beach resort could have devastating effects.
- Emergency services: richer countries usually have well-trained and well-resourced response teams that can rescue and treat people following a disaster.

Figure 1 Waves crashing against a sea wall.

Now test yourself

'People who live in coastal communities in low-income countries (LICs) are likely to be more vulnerable to coastal erosion and flooding than people who live in high-income countries (HICs).' To what extent do you agree with this statement?

TESTED

How severe weather events and climate change create vulnerability to coastal flooding

Over 1 billion people live in **low-elevation coastal zones (LECZs)**. Three-quarters of the world's mega-cities have a coastal location. Risks to coastal communities are likely to increase in the future because:

- Sea levels are likely to rise by between 50 and 100 cm by 2100, which will increase coastal flooding.
- Warmer seas lead to increased frequency and strength of storms causing increased coastal erosion and **storm surges**.
- More violent storms will give heavier rainfall and increased risk of flash flooding.

Economically developing countries are likely to be in the frontline of the impacts of climate change. These countries are the most vulnerable to the risks but have the least capacity to react and recover.

Low-elevation coastal zones (LECZs) Coastal areas that are less than 10 m above sea level

Storm surge A rapid rise in sea level caused by storms forcing water into a narrowing sea area such as an estuary

Isostatic change Change in the height of land relative to the sea, often because of the melting of ice from the last ice age

Example of a vulnerable coastal landscape: Thames Gateway

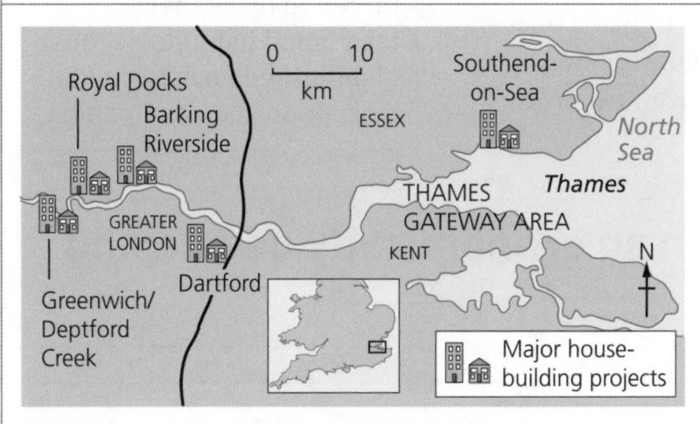

Figure 2 Location of the Thames Gateway.

One of the most vulnerable coastal landscapes in the UK is the Thames Estuary to the east of London. This stretch of coastline is known as the Thames Gateway. It is at risk of severe weather events. These risks include:

- storm surges: the sea is pushed into the funnel-shaped estuary
- **isostatic change**: the area is sinking into the sea by about 2 mm per year
- changing sea levels: the sea level is rising at about 3 mm a year.

The area is particularly vulnerable because:

- 1.6 million people live and work in the area.
- 500,000 properties are at risk.
- 75 per cent of the property value in England and Wales lies on the Thames Estuary.
- London City Airport is at risk.
- London is the UK's largest centre of economic activity, contributing £250 billion to the UK economy each year.

Storm surges are a particular threat because:

- Depressions over the North Sea create lower pressure causing a rise in sea level.
- Northerly winds push the surface waters of the sea forward, a motion known as 'wind drift'.
- The North Sea is shaped like a funnel and as water is forced southwards it cannot escape through the narrow English Channel; this increases the height of the sea.
- The funnelling effect is further enhanced by the Thames Estuary.
- Following devastating flooding in 1953, the government decided to build the Thames Barrier.

Revision activity

Carry out your own internet research. On a sheet of A5 card, build up a Five Ws case study of the storm surge which hit the east coast of Britain in December 2013. The Five Ws are questions which will help you to gather information or problem-solve. They will provide you with a summary of the winter storm.

- **What** happened?
- **When** did it take place?
- **Where** did it take place?
- **Why** did it happen?
- **Who** did it affect?

Each question should have a factual answer.

Now test yourself

1. Why are the risks to coastal communities likely to increase in the future?
2. What is a storm surge?

TESTED

Social and economic factors that increase vulnerability in countries at different levels of economic development

- Primary impacts are the immediate consequences of an event, such as people inside a collapsing building during a cliff collapse.
- Secondary impacts are those which result from the primary event, for example destroyed buildings lead to people becoming homeless. Secondary impacts may last for many years after the event.

In the UK, a high-income country, over 20 million people live within 10 km of the coast. Many coastal communities are particularly vulnerable because of factors such as high numbers of older residents, tourists and visitors who stay for a short time only, low employment levels, seasonality of work and poor transport links.

In economically developing countries, agriculture is often the backbone of their economies and millions of people live on fertile river **deltas** such as in Egypt. Many of the world's largest cities are located in the coastal areas of economically developing countries. Mumbai, in India, is one example where 18.4 million people live, many in poorly built housing on land less than 10 m above sea level in an area at risk from tropical cyclones.

Delta A landform created by the deposition of sediment carried by a river as the flow leaves its mouth and enters slower-moving or standing water, for example an ocean, sea or lake

Salination The process that increases the salt content of water or soil

Environmental refugees People who have been forced to leave their traditional habitats because of a marked environmental disruption such as flooding

Example of a coastal community in an economically developing country: Nile Delta, Egypt

The Nile Delta is one of the oldest intensely cultivated areas on Earth. Almost 40 million people live in the delta region, with population densities up to 1600 inhabitants per km^2. The low-lying, fertile floodplain of the Nile valley and the Nile Delta is surrounded by deserts.

Alexandria is the largest city in the delta, built in Egypt's LECZ. It has a population of more than 4.5 million – half of these people live in self-built informal housing. These urban poor are at great risk from a sea level rise because they:

- have few savings and no insurance
- rely on boreholes for drinking water, which is increasingly contaminated
- live in unhealthy conditions
- live in poorly built multi-storey buildings.

The 50 km wide strip of land along the coast is less than 2 m above sea level. It is protected from flooding by a sand belt that ranges from 1 to 10 km wide. Rising sea levels will destroy parts of this sand belt. The impacts would be serious:

- one-third of Egypt's fish catches are made in lagoons protected by the sand belt
- fertile agricultural land would be lost
- low-lying parts of Alexandria and Port Said would be destroyed
- beach resorts would be endangered and tourism threatened
- it will cause serious groundwater **salination**
- 8 million people may become **environmental refugees**.

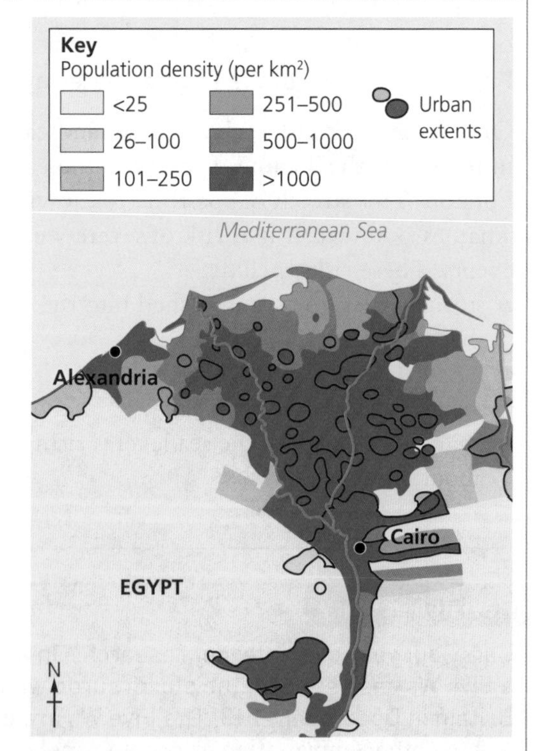

Figure 3 Population density of the Nile Delta.

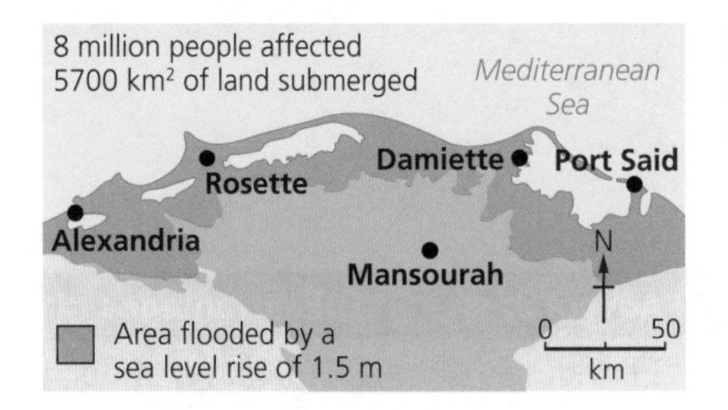

Figure 4 Impact of sea level rise on the Nile Delta.

Revision activity

On a sheet of A5 card complete a summary which highlights:

1. The **social** and **economic** factors which increase the vulnerability of people who live on the Nile Delta.
2. The impacts of sea level rise on the people who live on the Nile Delta.

Now test yourself

TESTED

1. Explain why the following factors increase the vulnerability of coastal communities:
   - high numbers of older residents
   - tourists and visitors who stay for a short time only
   - poor transport links.
2. Describe the possible impacts of rising sea levels on the people who live on the Nile Delta, Egypt.

Exam practice

Figure 5 Impact of Cyclone Phailin on the coast of the Bay of Bengal, India.

Study the photograph.

Use evidence in the photograph only to describe two ways in which the cyclone affected the lives of people who live in the coastal communities affected by Cyclone Phailin. [4]

Exam tip

You must only use evidence in the photograph. Make sure you develop points so that you gain all of the marks available in a question.

Managing coastal hazards

How are coastlines managed?

REVISED

Many different groups of people have an interest in what happens at the coast including:

- residents
- environmental groups
- developers
- local councils
- national governments
- tourist boards
- National Parks Authorities.

Each of these groups may have a different view about what should be done to manage the coastline. Management is needed because:

- Homes and businesses need protection from coastal erosion, landslides and flooding.
- Sea levels are rising, increasing the risk of coastal flooding.
- Litter, pollution from sewage discharge or accidents such as oil spills need to be cleaned up.
- The magnitude and frequency of storms is increasing, causing storm surges, flooding and wind damage.
- Natural habitats and heritage sites need preserving and protecting.

There are two main methods of defence used to protect the coastline: **hard engineering** and **soft engineering**.

Hard engineering Building an artificial structure to control coastal processes. It is usually expensive, has a high impact on the environment and is unsustainable

Soft engineering Works with natural systems, using natural materials and processes. It is often less expensive, has a low impact on the environment and is more sustainable

Hard engineering

Hard engineering uses structures or machinery to control coastal processes. It is an expensive and short-term option. It has a high impact on the landscape and environment and it is unsustainable.

Image	Method	Description	Advantages	Disadvantages
	Sea walls (£6000 per metre)	Concrete sea walls reflect the energy of waves and prevent flooding. They are often recurved which means waves are reflected back on themselves. They are often used to protect settlements	Provide excellent defence where wave energy is high. Have a long life span	Expensive, affect access to the beach, recurved sea walls can increase erosion of beach material
	Groynes (£5000 each)	Wooden barriers (usually), built down a beach, trap sand being transported by longshore drift. The resultant wider beach absorbs wave energy, reducing the rate of cliff erosion	Relatively cheap, retain a wide sandy beach	Beaches further along the coast are starved of beach material
	Rip rap (£1000 per metre)	Large boulders of hard rock are placed along the base of a cliff and absorb the energy of the waves	Relatively cheap and efficient	Unattractive, access to beach becomes difficult, costs increase when rock is imported
	Gabions (£100 per metre)	Steel cages, containing boulders, absorb the energy of the waves	Cheap and efficient	Visually unattractive. Shorter life span than a sea wall
	Revetments (£2000 per metre)	Traditionally these are wooden fence-like structures that allow seawater and sediment to pass through, but the structures absorb wave energy. A beach builds up behind the revetment and provides further protection	Cheaper and less intrusive than a sea wall. Cause less erosion of beach material	Short life span and unsuitable where wave energy is high

Revision activity

On five sticky notes draw and label a sketch of each method of sea defence. Label each sketch with two advantages and two disadvantages. Stick them where you can see them. Look at them every night and every morning until you can remember them.

Now test yourself TESTED

Explain why many people have the viewpoint that hard engineering is an unsustainable way of managing our coastline.

Soft engineering

Soft engineering involves working with nature. It is often less expensive than hard engineering, and is usually more sustainable with less impact on the environment. Examples include:

Method	Description	Advantages	Disadvantages
Beach nourishment	Beaches are made higher and wider by importing sand and shingle from further along the coast or dredging it from the sea bed	• It is relatively cheap, approximately £20 per m^3 • It retains the natural look of the beach	• Offshore dredging of sand and shingle increases erosion in other areas and affects ecosystems • Beach replenishment is necessary on a regular basis, increasing costs
Sand dune stabilisation	Sand dunes act as a natural defence against coastal flooding and erosion	• Dunes are left undisturbed, hence maintaining the natural ecosystem • Boardwalks are constructed, sections of sand dunes are marked as out of bounds, so the dunes become more accessible to tourists	• Management is time consuming, for example planting marram grass and fencing off areas • Cost is expensive, around £2000 per 100 m
Managed retreat (managed realignment)	Areas of coast are allowed to erode and flood naturally, creating a new **intertidal zone** which acts as a natural buffer against storms and rising sea levels. Usually used in areas where land is low value	• It retains the natural balance of the coastal system • Eroded material encourages the development of beaches and salt marshes	• The cost depends on the amount of compensation that needs to be paid to landowners and homeowners • People lose their livelihoods and homes

'Hold the line' and 'managed retreat'

There are contrasting approaches to coastal management:

- **Hold the line**: where existing coastal defences are maintained. Hard and soft engineering is used to keep the coastline in the same place. This is a popular option with local residents.
- Managed retreat (surrender): move people out of danger zones and let Mother Nature take control. The coastline will move inland, only being defended when necessary. This is a less expensive option but usually less popular.

Hold the line Existing coastal defences are repaired but no new defences are set up

Beach nourishment Sand and shingle are added to a beach to make it higher and wider

Sand dune stabilisation Planting vegetation, such as marram grass, or building wooden fences helps sand to build up and the dunes stabilise, which then provides a barrier and absorbs wave energy

Managed retreat Involves allowing the sea to breach existing defences and flood the land behind it

Intertidal zone The area that is above water at low tide and under water at high tide

Example of 'hold the line': Borth, Ceredigion

Figure 6 Wooden groynes on Borth Beach in 2009.

The village of Borth is built on the southern end of a pebble ridge that sticks out into the Dyfi Estuary. Wooden groynes, built in the 1970s, trapped sand on the beach which protected the village. By the 1990s these groynes were in poor condition. Ceredigion Council needed to decide on its future management strategy. Should the council 'hold the line' or carry out a 'managed retreat'? If it decided on a managed retreat then:

- Storm waves would flood Borth in the next 10–15 years.
- Property in Borth worth £10.75 million would be lost.
- The peat bog north-east of Borth, recognised by UNESCO and a Special Area of Conservation, would be flooded and its existing ecosystem lost.
- The nearby Ynyslas sand dunes which attract thousands of holidaymakers every year would be cut off from the town.
- Many local businesses would suffer if tourist numbers fell.

It was estimated that new coastal defences would cost £7 million. In 2000 a decision was made to 'hold the line'. In 2010 work began:

- Four rock groynes were built to trap sediment being moved along the beach by longshore drift and maintain a wide beach in front of the village. The rock for the groynes was purchased from the local quarry.
- An artificial rock reef was built parallel to the shore to break the power of the waves, hence reducing erosion and encouraging the deposition of sand and pebbles.
- Beach nourishment was put in place to increase the width of the beach so that waves will break further offshore.

The scheme was completed in 2015 at an actual cost of £18 million. The scheme so far has been successful in preventing further damage and flooding.

Revision activity

Find out where Borth is on the Welsh coastline. On sticky notes make a list of all of the different groups of people you think would be involved in making the decision to 'hold the line' at Borth. Write each 'group' on an individual note. Now move the sticky notes around to show the following:

- People who would have helped make the decision to protect the coastline and people who would not have helped.
- People positively and negatively affected by the decision to protect Borth.
- People affected by the decision in the short term and people affected in the long term.

We cannot afford to 'hold the line' by repairing sea defences. This strategy is not sustainable in the face of rising sea levels. We believe that 'managed retreating' is the only practical option. **Local councillor and Green Party member**

My family have farmed this land for years. It is fertile farmland and our country needs food, we can't import all of our needs. **Local farmer**

Sea levels are rising and it is not possible to continue to defend our entire coastline. In some areas people will have to be moved out and the sea allowed to flood what is naturally their domain. **Environment spokesperson in the national government**

This is our home and we don't want to leave it. The local council should protect us from coastal erosion by replacing sea defences. **Local retired resident**

The population of our country is growing and people need to be housed. Coastal areas provide some of the best building land in the country in a very desirable location, where people want to live. **Property developer**

Now test yourself

TESTED

1. Describe the differences between 'hold the line' and 'managed retreat' coastal management strategies.
2. Explain why two different groups of people may have different views about what should be done to manage a stretch of coastline.
3. If you were responsible for devising a management strategy for a stretch of coastline, how would you take into consideration the views of all of the different groups of people shown above?

Exam tip

It is important to focus on the command word when answering a question. The two most commonly used command words are **describe** and **explain**. Describe invites you to 'paint a picture' of what something is like by using lots of adjectives. Explain directs you to say why in this case people have different views.

The cost-benefit of coastal protection

Cost-benefit analysis involves adding up the benefits of a course of action, and then comparing these with the costs associated with it. Where the cost of managing a coastline would be greater than the benefits likely to be gained from protecting it, particularly in areas where there are few people and little of 'value' to protect, the decision may be to do nothing and allow flooding and erosion to happen.

Cost-benefit An analytical tool for assessing the pros and cons of a decision

Social and economic reasons why some coastlines are protected

Failure to manage coastlines will have severe economic and social effects along coastlines used for settlement, tourism and industry. Management of coastlines is also important to help protect natural habitats. Governments generally do not engage in coastal management where there is not an economic risk. Effective coastal management is expensive and questions are increasingly being asked as to whether it is worth the money.

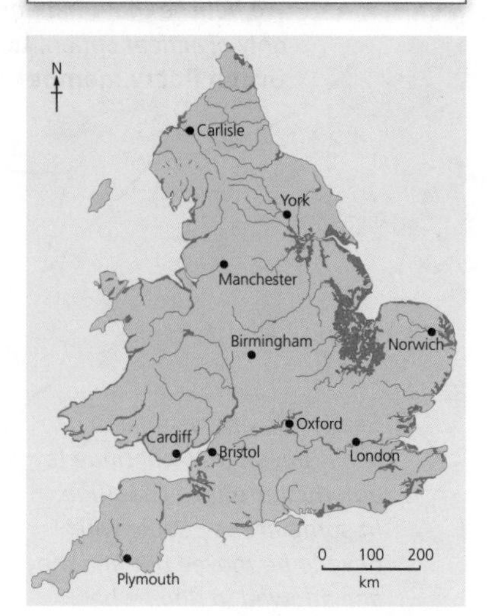

Figure 7 Flood zones in England and Wales.

Example of 'managed retreat': Medmerry

In 2014, the UK's largest realignment sea defence scheme was completed in Medmerry, West Sussex. Problems of flooding from the sea had affected the area for many years and up to £300,000 was spent every year repairing the shingle bank protecting the area.

The new defence scheme has redrawn the coastline 2 km further inland. It involved the construction of 7 km of earth embankments, four channels to take freshwater through the defences and a breach in the existing sea wall to allow land to flood.

The completed scheme:

- cost £28 million
- protected 348 properties, the sewage works and the main road into Selsey serving over 5000 households
- created 183 ha of intertidal habitat, managed by the RSPB as a nature reserve
- saved £300,000 a year in maintenance costs
- allowed public access and a boost for green tourism
- allowed cattle to be grazed on the salt marsh, producing highly valued beef
- provided improved public health with access to the countryside and wildlife.

Has the scheme worked?

- The scheme performed well during the first year of operation. No flooding issues were experienced despite the extreme weather in the winter of 2013–14.
- The habitat is developing well, with good populations of birds and other wildlife using the site.
- The scheme is delivering benefits to the local economy. The nature reserve is attracting more visitors to the area, and the large local caravan park has extended its opening periods by two months, generating income and job security.

Now test yourself

TESTED

1. Using evidence from the map, describe the distribution of flood zones in England and Wales.
2. 'Governments generally do not engage in coastal management where there isn't an economic risk.' What is meant by this statement?
3. Identify one group of people who are likely to be in favour of the Medmerry scheme and one group who would be against. Explain their viewpoints.
4. Carry out a cost-benefit analysis of the Medmerry scheme and decide whether in your opinion it has been successful.

Shoreline management plans (SMPs)

The management of coastal areas in the UK is covered by **shoreline management plans (SMPs)**. It is the responsibility of local councils to develop SMPs. Each should consider:

- How many people are threatened by erosion and flooding and how much would it cost to rehouse these people?
- How much would it cost to rebuild roads and railway lines if they were washed away?
- Are there historic or natural features that should be conserved and do these features have any economic value, for example as tourist attractions?

Shoreline management plan (SMP) An assessment of the risks associated with coastal processes

Cliff regrading Reducing the angle of a cliff to reduce mass movement

Example of management strategies: Holderness Coast, Yorkshire

The Holderness Coast in Yorkshire has some of the highest erosion rates in Europe:

- Cliffs erode at an average of 2 m a year.
- Rocks are mostly weak boulder clay.
- Erosion rates are highest when spring tides combine with winds from the north-east.
- Beaches are narrow and offer little protection.
- The coastline has retreated 4 km since Roman times and many villages have been lost.

Risks include:

- Fertile farmland will be lost to the sea.
- More villages will disappear.
- Major roads will disappear into the sea.

Management strategies:

- Hornsea has been protected with groynes, a sea wall and rip rap.
- Mappleton has been protected with groynes, rip rap and **cliff regrading**.

These sea defences have increased erosion further south at Cowden and Aldborough. Groynes have trapped sand and starved the beaches downdrift, exposing cliffs to the full force of the waves.

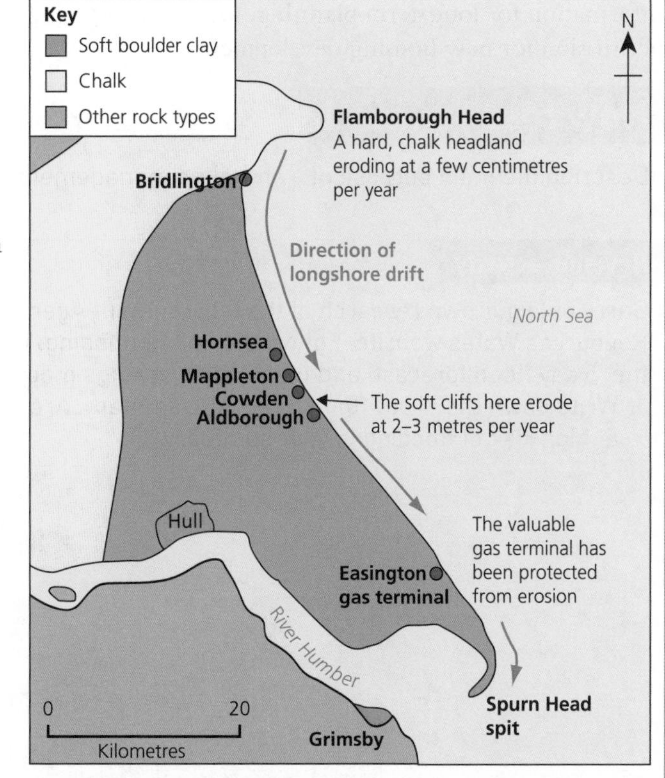

Figure 8 The Holderness Coast.

Co-ordinating coastal management

The Holderness Coast is an example of where protecting the coastline in one area can cause problems elsewhere. It demonstrates the need to co-ordinate coastal management at a regional and national level.

Integrated coastal zone management (ICZM) is a concept born in 1992 during the Earth Summit of Rio de Janeiro. It aims to develop sustainable solutions based on:

- Which areas need protecting and which areas are not cost effective to protect?
- What type of defence should be used?
- What is best for wildlife and the natural environment?
- What is the best solution for the people who live at the coastline, taking a balanced view and not favouring one group over another?

Monitoring, hazard mapping and emergency planning

The capacity of a community to react to and recover from a coastal hazard depends on how prepared that community is for that hazard. This depends on:

- Monitoring: being able to predict extreme weather.
- Hazard mapping: being able to know which areas will be affected by the hazard.
- Emergency services: having the resources and training to react to the hazard.

In the UK, the Flood Forecasting Centre monitors the weather and makes flooding forecasts. The Environment Agency and Natural Resources Wales inform the public of these forecasts and also produce **hazard maps** to identify areas at particular risk. A hazard map alerts the public to the areas at danger of coastal flooding and gives local authorities information for long-term planning, for example giving planning permission for new housing developments.

Hazard map A map that highlights areas that are affected by, or vulnerable to, a particular hazard

Now test yourself

TESTED

Describe the main purpose of a shoreline management plan (SMP).

Revision activity

Carry out your own research of the Environment Agency or Natural Resources Wales website. Follow the link to flooding. Now research the '5 day flood forecast' and the 'Flood warnings map' links.

1 Write down the '5 day flood forecast' summary. Are there any areas at high risk of flooding in your summary?

Figure 9 **Screenshot of the Natural Resources Wales flood hazard map for Kinmel Bay and West Rhyl.**

2 Below the flood warnings map you will find a link to 'What to do before a flood' and 'What to do during or after the flood'. Imagine that a forecast has predicted that your area is at high risk of flooding. Produce a 60-second radio alert informing the public about what to do in the event of a flood.

3 Draw a labelled sketch map to show areas of Towyn and Rhyl that are at risk of flooding.

4 Below your map give two ways in which this hazard map would be useful in reducing the risks to people of coastal flooding.

Exam practice

Describe how hard engineering strategies may be used to reduce the risk of coastal erosion and flooding in one location you have studied. [4]

Exam tip

It is important to use examples in your answers to exam questions. In a points marked question these will give you extra marks. In a levels marked question they will encourage the examiner to give you a higher level. Often in a levels marked question a lack of examples will mean you cannot achieve full marks.

What is the most sustainable way to manage coastlines in the face of rising sea levels?

REVISED

A spokesperson for the **Environment Agency** says 'It is important that we now consider a 100-year timeframe which forces us to look at the ways in which our coast will change and consider the impact that climate change will have.'

Environment Agency A non-departmental public body with responsibility for the protection and enhancement of the environment in England (and until 2013 also Wales)

Tropical storm An intense low pressure weather system that can last for days to weeks within the tropical regions of Earth. Known as hurricanes in North America, cyclones in India and typhoons in Japan and East Asia

Reasons for increased vulnerability of coastal communities in the future

Climate change is placing increasing pressure on coastal regions which are already seriously affected by intensive human activity. This raises the question of whether people can continue to live in low-lying coastal regions. Coastal ecosystems and habitats such as mangroves, coral reefs, seagrass meadows and salt marshes are also under threat.

- More than 200 million people worldwide live along coastlines less than 5 m above sea level.
- By the end of the century, this figure is estimated to increase to 400–500 million.
- These areas could be inundated by rising sea levels which are forecast to rise by up to 1 m.
- Inhabitants will be forced to find ways of managing rising water levels or to abandon some areas altogether.
- Governments cannot defend everything at all costs.

Scientists also believe that climate change could lead to increased coastal erosion, due to more aggressive marine conditions, and to an increase in the magnitude and frequency of storms. This is a particular concern for those people who live in regions of the world affected by **tropical storms** and the devastation they can bring.

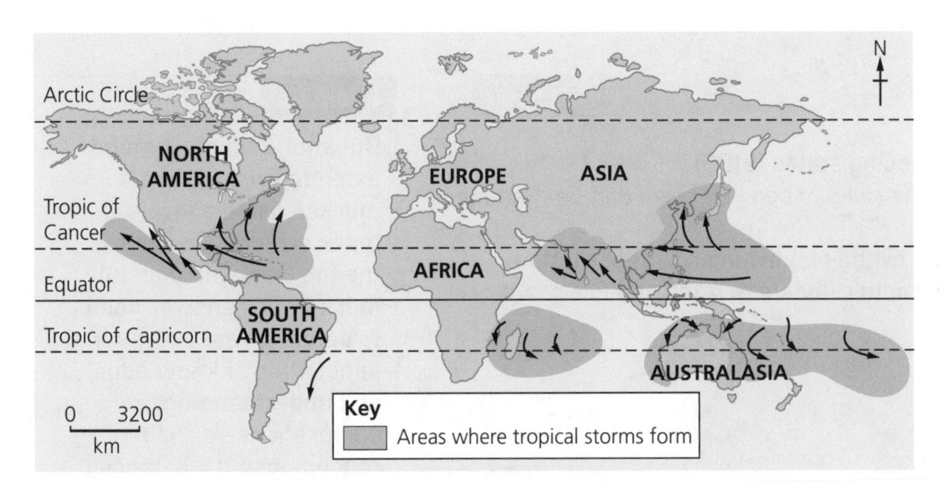

Figure 10 Tropical storm tracks and the areas where they form.

Revision activity

1. What different names are given to tropical storms in different parts of the world?
2. Carry out internet research and list ten countries affected by tropical storms.

Why some coastlines are at greater risk than others

As well as the general rise in sea levels, there are local factors that mean some coastlines are more at risk than others:

- Some coasts are sinking or subsiding. **River estuaries** and deltas sink under their own weight as more sediment is deposited. Parts of the city of New Orleans in the USA are subsiding by 28 mm per year. Northern parts of the UK were covered by thick layers of heavy ice during the last ice age. When the ice melted the land in this part of the UK began to rise slowly and the southern part of the UK began to sink.
- Rocks which make the coast may be hard or soft. Clay rocks of the Holderness Coastline give it some of the highest erosion rates in the world.
- Coastal storms affect some coastlines more than others. Tropical storms only affect coastlines in some parts of the world (see Figure 12, page 80). Some coastlines such as the Thames Gateway are more vulnerable to storm surges.
- Some coastlines are threatened by tsunami. In December 2004, the Indian Ocean tsunami affected thirteen countries and killed over 230,000 people.

Perhaps the communities that are at greatest risk are those living on the world's river deltas. Millions live on deltas in Bangladesh, Egypt, Nigeria, Vietnam and Cambodia.

Rising sea level challenges faced by small island states

Small island developing states (SIDS) are low-lying coastal countries first recognised as a distinct group in 1992. They share similar challenges, including small but growing populations, limited resources, remoteness and fragile environments.

Now test yourself

TESTED

1. What are small island developing states (SIDS)?
2. Give two examples of how the risks of sea level rise can be managed in the Maldives.
3. Do you think the suggestion by the Maldivian president that the population of the Maldives could relocate to Australia is a practical one?

Exam practice

'Small island developing states are more vulnerable to coastal hazards than any other location.' Do you agree with this statement? Explain your answer. [8]

Now test yourself

1. Give two reasons to explain why some coastal communities are at greater risk than others.
2. Give two reasons to explain why these communities will be more vulnerable in the future.

TESTED

River estuary The wide mouth of a river where it meets the sea

Small island developing states (SIDs) Low-lying coastal countries that tend to share similar development challenges, including small but growing populations, limited resources, remoteness, susceptibility to natural disasters, vulnerability to external shocks, excessive dependence on international trade and fragile environments

Exam tip

Questions which demand extended writing are marked using a levels mark scheme. To achieve the top level and gain full marks, your answer needs to demonstrate exceptional application of knowledge and understanding, a comprehensive chain of reasoning and a balanced appraisal.

Revision activity

Complete a wheel of knowledge for the topics in this theme (see www.hoddereducation.co.uk/myrevisionnotes).

Example of an SIDS and sea level rise: the Maldives

The Maldives is a nation of 26 coral islands located in the Indian Ocean. It has a population of 350,000. Eighty per cent of the land area is less than 1 m above sea level and nowhere is more than 3 m above sea level.

The Maldives is a poor country, ranked 165th out of 192 nation states. The main occupation is fishing although tourism has become increasingly important in recent years. The Maldives is ranked the third most endangered nation due to flooding from climate change. Risks include:

- Rising sea levels which threaten the existence of the Maldives. The islands may disappear in the next century. Inhabitants will become **environmental refugees**.
- Flooding of beach resorts, damaging the tourism that the economy depends on.
- More intense tropical storms.
- Damage to coral reefs due to warmer sea temperatures. Ecosystems associated with reefs are lost. This will affect diving tourism. Ninety per cent of government tax revenue comes from tourism.
- Money will be spent on sea defences at the expense of public services and development.
- Groundwater sources are increasingly contaminated by saltwater. Freshwater is scarce; 87 per cent comes from rainwater.
- Agricultural land is being lost and the fishing industry is disrupted.

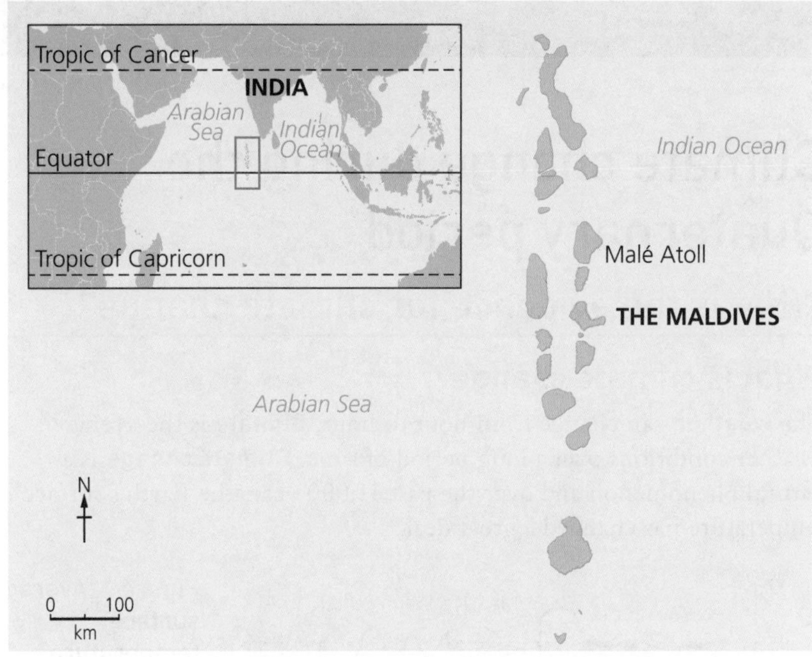

Figure 11 The location of the Maldives.

How can the risks be managed?

- A 3 m high wall has been built around Malé, the capital, at a cost of $63 million, funded by Japan.
- The smaller islands could be evacuated. This is a temporary solution because eventually all the islands would also have to be abandoned. The United Nations predicts that the Maldives will be uninhabitable by 2100.
- **Dykes** could be built to hold back the sea. The cost of building a dyke depends on its length, so this is a very expensive option in an island state.
- The height of the islands could be increased; this will require a lot of sand and coral.
- The Maldivian president has suggested that the population could be relocated to Australia.

Revision activity

1 Consider the impact of sea level rise on the Maldives and complete a table such as the one below:

Impacts	Social	Economic and environmental
Primary	(Impacts on the community and well-being of individuals and families)	(Impacts on wealth and ecosystems)
Secondary		

2 Use a mind mapping technique to summarise how coastal hazards can be managed.

Environmental refugees People who are forced to leave their home region due to changes in their local environment which could include drought, desertification and sea level rise

Dyke An artificial earthen wall built to prevent flooding by the sea

Theme 5 Weather, Climate and Ecosystems

Climate change during the Quaternary period

What is the evidence for climate change?

REVISED

What is climate change?

The **weather** can change from hour to hour. **Climate** is the average weather conditions over a long period of time. **Climate change** is a natural phenomenon and over the past 11,000 years the Earth's surface temperature has changed a great deal.

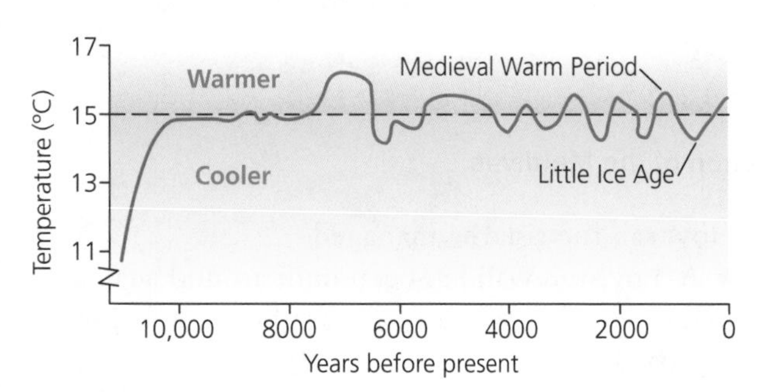

Figure 1 Average surface temperature changes of the northern hemisphere during the past 11,000 years.

During the Medieval Warm Period ice-free seas allowed the Vikings to colonise Greenland. The Little Ice Age was a colder period in northern Europe where temperatures were on average 1–1.5 °C colder than today.

Over a longer period of time, the past 400,000 years, there have been natural cycles of cooling and warming. The periods of cool temperatures, where average global temperatures are below 15 °C, are known as glacials and periods of warmth are known as interglacials.

Weather The atmospheric conditions at a particular place and time; includes temperature, precipitation, wind and sunshine

Climate The average weather over a long period (at least 30 years) of time

Climate change A large-scale, long-term shift in the Earth's weather patterns, especially in average temperatures

Evidence The body of facts or information which indicates whether a belief or theory is true

Carbon dioxide (CO_2) A colourless and odourless gas, made of carbon and oxygen, found in the atmosphere

Glaciation The process by which the land is covered by glaciers

Evidence for climate change

Evidence that climate has changed in the past includes:

- Fossils of plants and animals found in places where they could not live today.

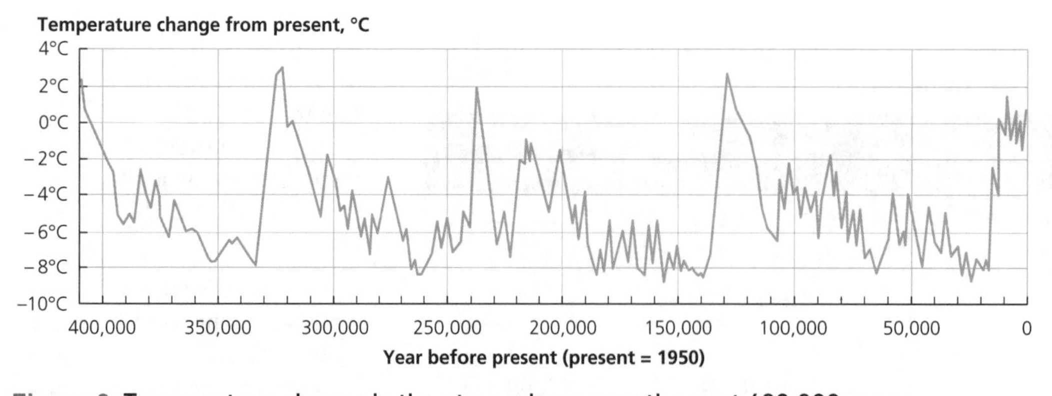

Narrow peaks in the temperature record represent short warm episodes (interglacials)

Broad dips in temperature represent glacial periods

Figure 2 Temperature change in the atmosphere over the past 420,000 years.

- Ice cores from the Antarctic show that the amounts of **carbon dioxide (CO_2)** and methane in the atmosphere have changed over the past 420,000 years.
- **Glaciation** in places now free of ice.
- Studies of tree rings, known as dendrochronology, show that growing seasons have varied in length in the past.
- Historical records such as diary extracts, crop yields for local registers and paintings, such as of ice fairs on the River Thames during the Little Ice Age.

Recent evidence of climate change includes:

- Increasing levels of CO_2 in the atmosphere.
- Shifting seasons leading to changes in the migration patterns of birds and insects.
- Glaciers and ice sheets melting and retreating.
- Measurements by the Met Office show that average global temperatures have increased by 0.6 °C in the past 100 years.

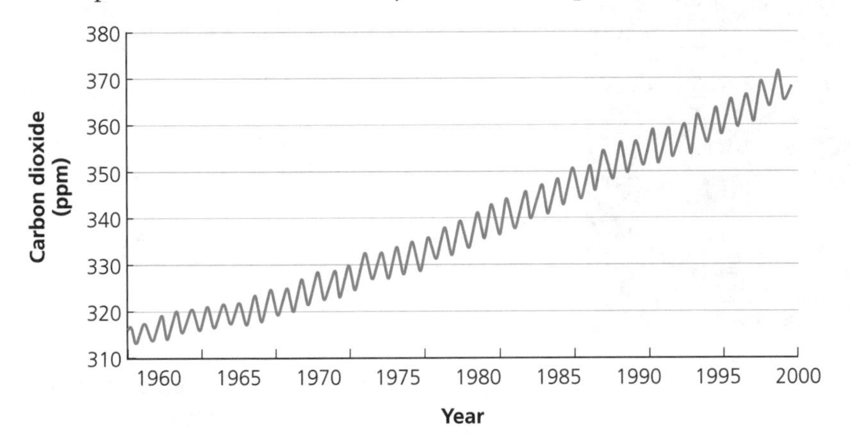

Figure 4 The Keeling curve plots the change in the amount of CO_2 in the Earth's atmosphere. Measurements started in 1958.

Example of the Pasterze Glacier

The Pasterze Glacier in Austria has retreated by about 8 km in the past 160 years.

Figure 3 The Pasterze Glacier photographed in 2015 showing its retreat since 1980.

Now test yourself

TESTED

Look at the Keeling curve and answer these questions:

1. How much CO_2 was in the atmosphere in 1960?
2. Describe the change in the amount of CO_2 in the atmosphere between 1960 and 2000.
3. Explain the annual cycle of CO_2 in the Earth's atmosphere.

Exam practice

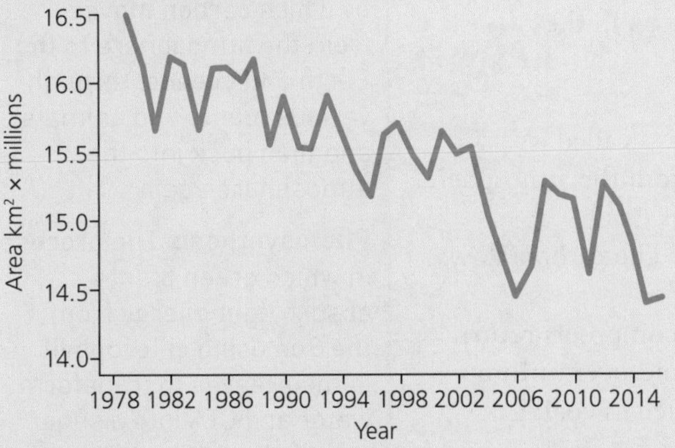

Area km² × millions: 16.5, 16.0, 15.5, 15.0, 14.5, 14.0
Year: 1978 1982 1986 1990 1994 1998 2002 2006 2010 2014

Figure 5 Average monthly Arctic sea ice extent, March 1979–2016.

Describe the change in the extent of Arctic sea ice between 1979 and 2014. [4]

Revision activity

On a sheet of A5 card, complete a spider diagram summarising the evidence for climate change. Include drawings to help you to remember each point.

Exam tip

If a question asks for a description of change in a graph, use figures to back up your description.

What are the causes of climate change?

REVISED

Scientists believe natural climate change is caused by:
- changes in the Earth's orbit
- changes in the tilt of the Earth
- changes in the output of solar radiation
- volcanic activity.

Example: Milankovitch cycles

The Serbian astronomer Milutin Milankovitch explained long-term climate change through changes in the Earth's orbit and rotation, known as Milankovitch cycles.

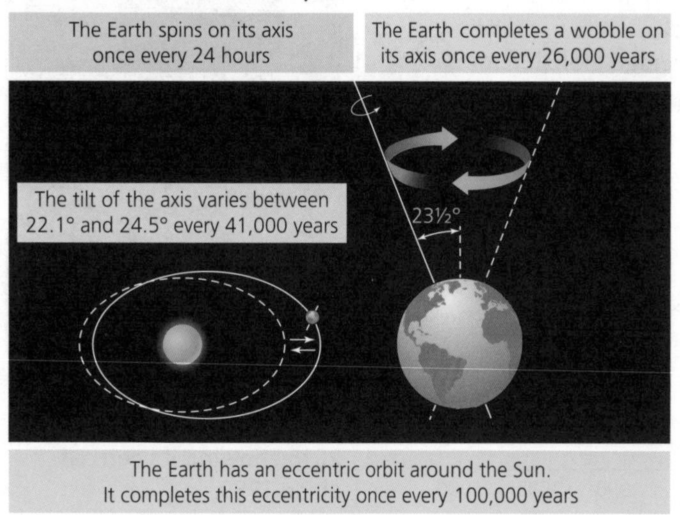

Figure 6 The Milankovitch cycles.

Warmer and cooler periods are caused by:
- The orbit of the Earth: it is sometimes closer and sometimes further away from the Sun.
- The tilt of the Earth: affects the amount of energy it receives from the Sun.

Now test yourself

1. Describe how the orbit of the Earth changes.
2. Explain why changes in the Earth's tilt could cause changes in the amount of energy received from the Sun.
3. Explain why large-scale destruction of forest areas will increase the amount of CO_2 in the atmosphere.
4. Explain why carbon is transferred more rapidly between some stores.

TESTED

Carbon flows The movement of carbon between stores in the carbon cycle

Carbon stores In the short term, carbon is kept or stored in the atmosphere, oceans and biosphere; in the long term, carbon is stored in fossil fuels

The carbon cycle process

REVISED

What is the carbon cycle?

All living things are made of carbon. Carbon is also a part of the ocean, air and rocks. Carbon moves or **flows** between the **stores** in the environment through a process called the **carbon cycle**:
- In the atmosphere, carbon is stored as CO_2.
- Plants use CO_2 and sunlight to make food by a process called **photosynthesis** during the day. The carbon flows from the atmosphere and is stored in the plant.
- Plants and animals give out CO_2 during respiration. The carbon flows back to the atmosphere.
- When plants and animals die, carbon is recycled, decomposers return it to the atmosphere as CO_2 or they may be buried and over millions of years turned into long-term stores as fossil fuels such as coal, gas and oil.

Carbon cycle The process by which carbon moves from the atmosphere to the Earth and oceans, through various plants and animals, and then back into the atmosphere again

Photosynthesis The process in which green plants absorb light energy from the Sun using chlorophyll in their leaves to transform water and CO_2 into a sugar called glucose

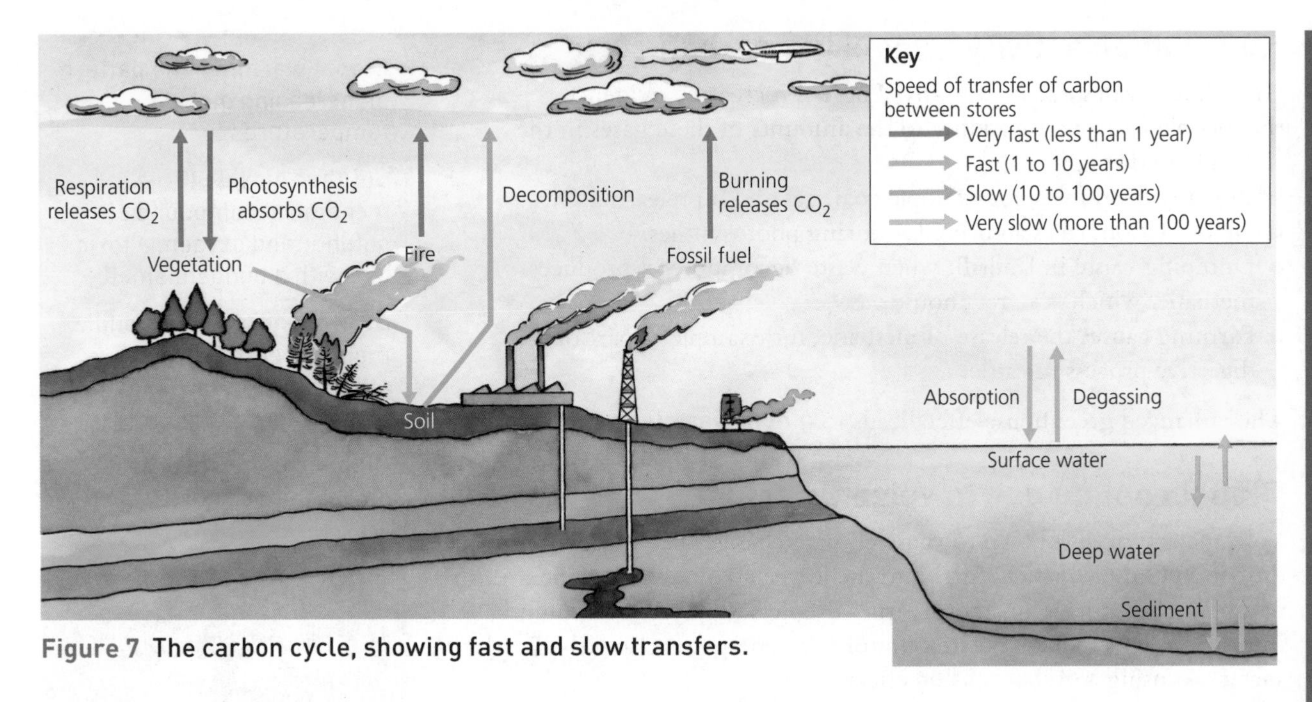

Figure 7 The carbon cycle, showing fast and slow transfers.

- When fossil fuels are burned the carbon is released and flows back to the atmosphere as CO_2.
- There is about 30 per cent more CO_2 in the air today than there was 150 years ago.

What is the greenhouse effect?

The **greenhouse effect** is the natural process by which the atmosphere warms up. Without it life would not be possible on Earth.

1. Solar energy enters the atmosphere.
2. As this short-wave energy passes through the atmosphere it may hit dust particles or water droplets and be scattered or reflected.
3. Only a little short-wave radiation is absorbed in the atmosphere.
4. Solar energy heats the Earth's surface, which then radiates long-wave (heat) energy into the atmosphere.
5. Long-wave energy is quite easily absorbed by naturally occurring greenhouse gases in the atmosphere. Of these, carbon dioxide is by far the most abundant.
6. Some long-wave energy escapes into space.

Key
Short-wave energy
Long-wave energy

Figure 8 The greenhouse effect.

Greenhouse effect
The natural process that results in the warming of the Earth's atmosphere

Now test yourself

TESTED

1. Cover the diagram above showing the greenhouse effect. Explain in your own words how the natural greenhouse effect warms the Earth's atmosphere.
2. Explain why the greenhouse effect is vital to life on Earth.

Revision activity

1. Complete a set of flashcards with questions and answers on the causes of climate change, the carbon cycle and the greenhouse effect. Test yourself using these cards.
2. Complete a spider diagram on a piece of A5 card to add to your set. Use the information in Figure 7 and your own research to show how the actions of people are increasing the level of greenhouse gases in the atmosphere, for example cattle release methane as they digest their food.

How human activity affects the carbon cycle

Greenhouse gases occur naturally. There is increasing evidence that people's actions are adding to the amounts of these gases in the atmosphere through:

- Burning fossil fuels, for example coal, gas and oil releasing CO_2.
- Deforestation: trees absorb CO_2 during photosynthesis.
- Dumping waste in landfill: when waste decomposes it produces methane, which is a greenhouse gas.
- Farming causes the release of methane, for example as part of the digestive process in cattle.

The enhanced greenhouse effect leads to **global warming**.

Global warming The pattern of increasing global temperatures

Sulphur dioxide (SO_2) A chemical compound of sulphur and oxygen, a toxic gas with a pungent smell

Stratosphere The second major layer of Earth's atmosphere, above the troposphere

Global cooling A decrease in global temperatures

Global cooling due to volcanic activity

Volcanic activity is known to cause climate change. Large eruptions eject dust and **sulphur dioxide (SO_2)** into the lower **stratosphere**. The mixture of ash and SO_2 forms an aerosol – tiny droplets that scatter sunlight back into space. This reduces the amount of solar energy reaching the Earth's surface, causing a **global cooling** effect.

Example: Mount Pinatubo

On 15 June 1991, Mount Pinatubo in the Philippines erupted:

- 10 km^3 of ash was ejected, blocking solar radiation.
- 15 million tonnes of SO_2 went into the stratosphere and formed a layer of sulphuric acid droplets which absorbed and scattered solar radiation.
- Mean world temperatures decreased by 0.5 °C.

Figure 9 Mount Pinatubo erupting.

Exam practice

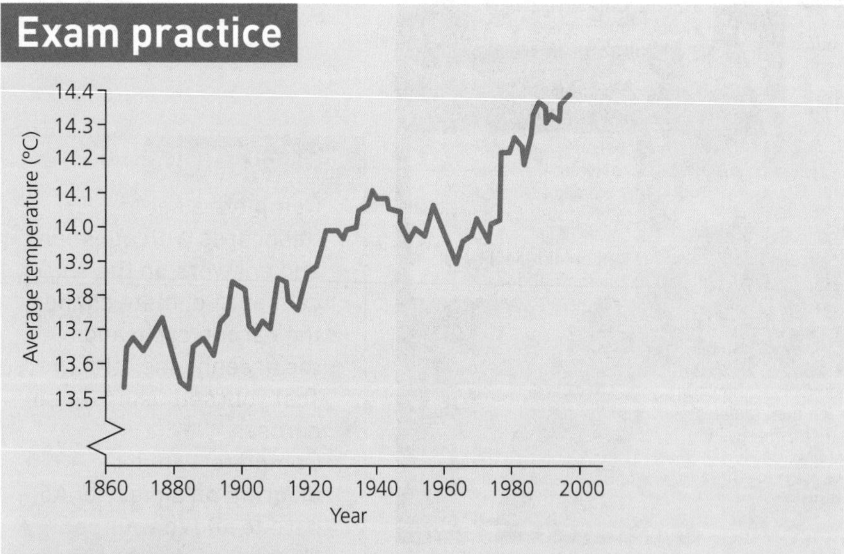

Figure 10 Global temperature changes, 1880–2015 (source: GISS/NASA).

1. Calculate the rise in the global average temperature between 1880 and 2000. [2]
2. Describe how one piece of evidence, other than rising temperatures, suggests that climate is changing. [4]
3. 'People's actions are the major cause of recent climate change.' To what extent do you agree with this statement? [8]

Exam tip

Questions worth less than 6 marks are usually marked using a point marking scheme. You will score 1 mark for every relevant point that is made. Look to develop points to gain full marks and use a 'chain of reasoning' to build an answer.

Now test yourself

How are human activities affecting the carbon cycle?

TESTED

Weather patterns and processes

What are the causes and consequences of weather hazards?

REVISED

Global circulation of the atmosphere

The global circulation is a worldwide system of winds which transports heat from tropical to polar **latitudes**:

- At the equator **insolation** heats the Earth which in turn heats the air above.
- Hot air rises, creating **low pressure**. When the air reaches the **tropopause** it cannot go any further and travels north and south.
- This air becomes colder and heavier, and at around 30° north and south it falls, creating **high pressure**.
- Air from the north and south then returns to the equator and meets in an area known as the **intertropical convergence zone (ITCZ)**.
- A large circulation of air is thus created, known as the **Hadley cell**.
- Air rises again at around 60° north and south and descends again around 90° north and south, creating a further two less distinct cells: the **Ferrel cell** and the **Polar cell**.

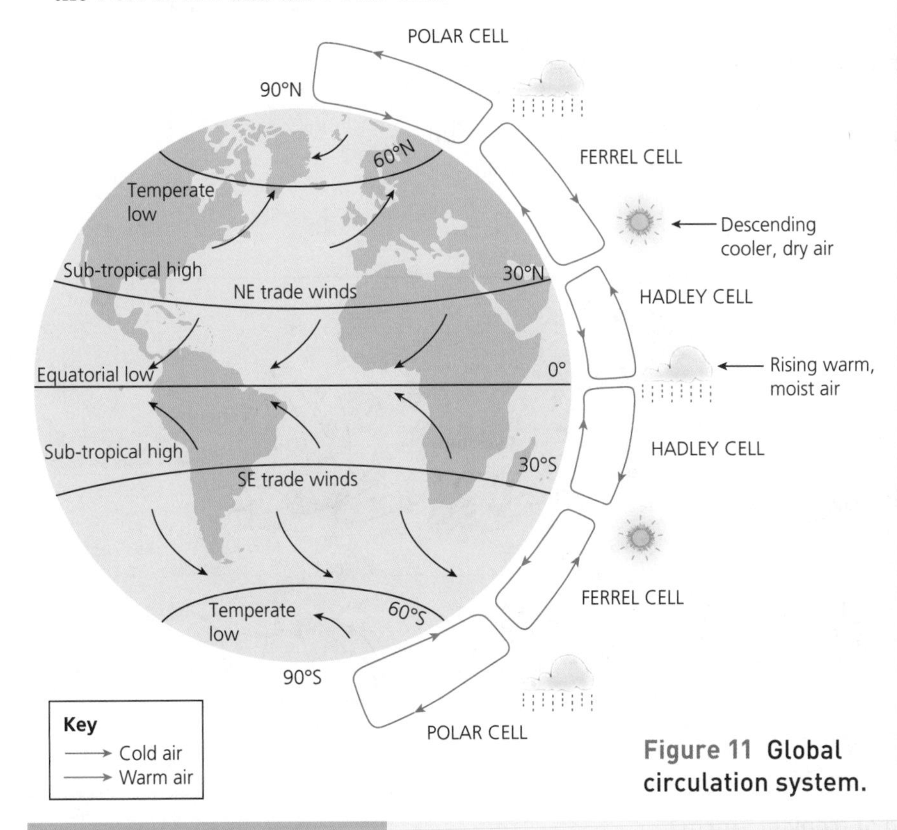

Figure 11 Global circulation system.

Latitude A measure of position north or south of the equator

Insolation Solar radiation that reaches the Earth's surface (energy received per cm^2 per minute)

Low pressure Rising air leads to low pressure at the Earth's surface

Tropopause The boundary separating the troposphere, where all weather takes place, from the stratosphere

High pressure Descending air leads to high pressure at the Earth's surface

Intertropical convergence zone (ITCZ) A zone of convergence at the equator where the trade winds meet

Hadley cell The portion of the tricellular model of air circulation where air rises at the equator due to convection, spreads in the upper troposphere and then sinks over the tropics before returning to the equator

Ferrel cell The mid-latitude cell in the tricellular model of atmospheric circulation

Polar cell A cell in the tricellular model of atmospheric circulation

Now test yourself

TESTED

Read the information on the global circulation of the atmosphere. Cover up the information once you have read it. Now decide which of the following statements are true:

1. Immediately to the north and south of the equator is a cell known as the Ferrel cell.
2. Air rises at the equator creating low pressure.
3. Winds from the north and south meet at the equator in an area known as the intertropical divergence zone.
4. Air rises at around 30° north and south of the equator creating high pressure.
5. There are a total of three large circulations of air north of the equator.

Changing patterns of tropical storms (low-pressure systems) over time

Globally, 80–100 **tropical storms** form over tropical oceans every year. There is an annual seasonality to their development. In the northern hemisphere, most tropical storms occur between June and November with a peak in September. In the southern hemisphere, the season lasts from November to April.

Evidence for long-term changes in the frequency and magnitude of tropical storms, as a result of climate change, is the subject of ongoing research. Some evidence suggests a recent increase in tropical storm activity. Others argue that variations in activity – on a regional and time scale – are part of natural cycles.

Tropical storm A severe low-pressure weather system which develops over tropical maritime areas

Storm surge The pushing of water against a coastline to abnormally high levels, usually a combination of extreme low pressure and winds pushing water into a narrowing feature such as a bay or estuary

How low-pressure systems lead to weather hazards

Tropical low-pressure systems are among the most powerful and destructive weather systems on Earth. Tropical storms (low-pressure systems) are known as hurricanes, typhoons or cyclones in different parts of the world.

Tropical storms form over tropical seas, when sea temperatures are over 27 °C. They move westward, can travel 600 km a day and have wind speeds that can reach over 120 km an hour. They give heavy rainfall, which can cause severe flooding, and die out when they reach land. High winds and low pressure generate large waves and **storm surges** which can cause flooding in coastal areas.

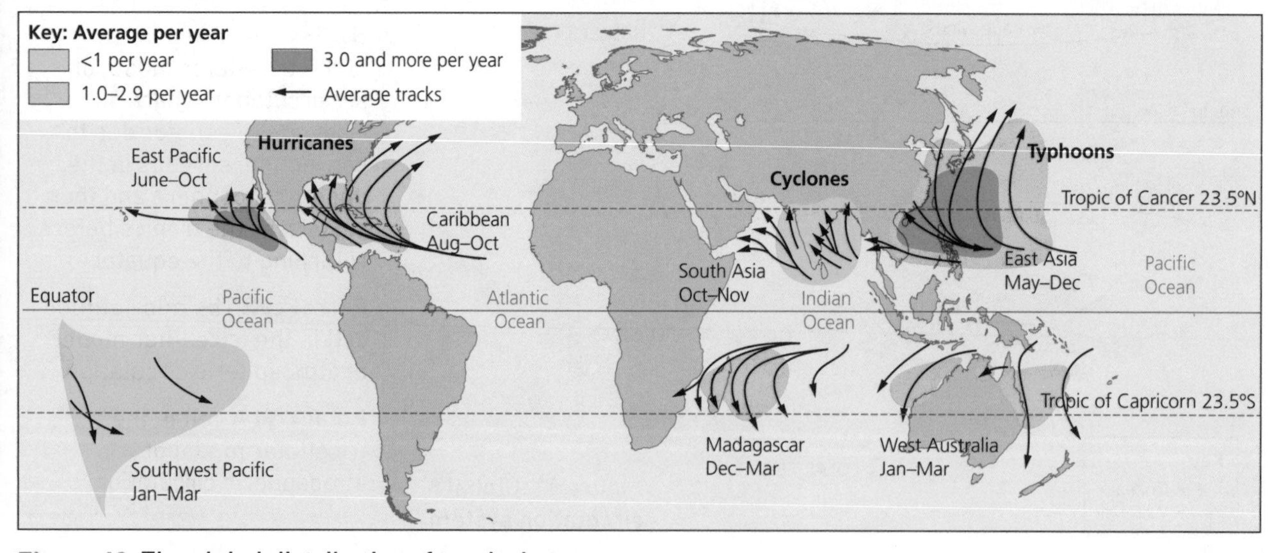

Figure 12 **The global distribution of tropical storms.**

Now test yourself

TESTED

1. What general name is given to tropical storms that affect places in the Caribbean?
2. Describe the distribution of areas of the world that are affected by tropical storms.
3. Explain why the majority of tropical storms occur between June and November in the northern hemisphere and from November to April in the southern hemisphere.

Example of a low-pressure hazard: tropical cyclone Pam

Tropical cyclone Pam hit the island chain of Vanuatu in the South Pacific in March 2015. Pam was a category 5 cyclone where wind speeds reached over 250 km an hour.

Social, environmental and economic impacts	Responses
Eleven people died	Emergency aid was sent by Australia, Fiji, France, New Zealand and the UK
90,000 people were made homeless	Repairs provided safe drinking water
Winds destroyed hospitals and schools	Blankets were distributed to the homeless to keep people warm
80 per cent of **subsistence** crops were flattened	153 temporary schools were set up
Coastal areas were flooded by a storm surge; freshwater wells were contaminated by seawater	Foreign medical teams arrived; 19,000 children were vaccinated against measles

Example: the South Asia monsoon

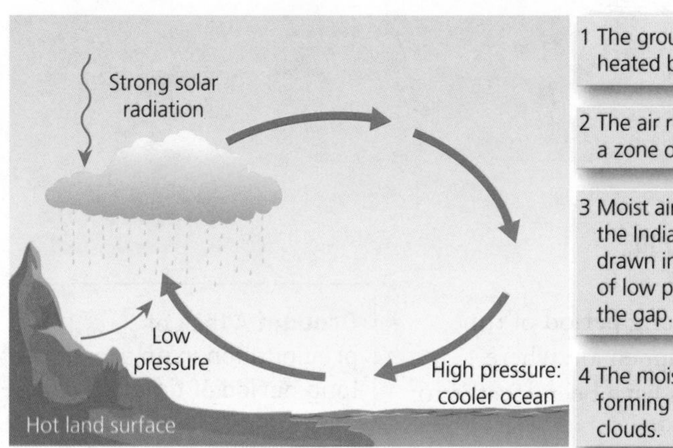

Figure 13 Circulation of the atmosphere over South Asia during July.

In South Asia the impacts tropical storms can be heightened by the monsoon climate. In July 2015, more than 200 people lost their lives and more than 1 million were made homeless when cyclone Komen hit the region.

A monsoon climate is characterised by a seasonal change in the direction of the **prevailing winds** which leads to distinct wet and dry seasons. In South Asia, the ITCZ moves northwards across India during July. Low pressure develops over Asia which pulls air in from the Indian Ocean, leading to heavy rains. During the northern hemisphere's winter, a large area of high pressure builds over Asia, pushing cool, dry air south, providing the region with its dry season.

Subsistence A farming system where farmers produce just enough to sustain themselves and their families

Prevailing wind The direction from which wind most frequently blows in a particular place

Revision activity

Imagine you live in an area affected by hurricanes. Devise a 'Hurricane Action Plan'.

- Discuss the type of hazards that could affect your family.
- Locate the safest places in your home or community for each hazard.
- Decide on a place to meet and actions you need to take once the storm has passed.
- Make up a slogan for your local council that will help people.

Global distribution of areas affected by heatwaves and drought

A **heatwave** is an extended period of hot weather relative to the expected conditions at that time of year. It is usually associated with high pressure which has become stationary over an area. Met Office research shows that in Europe, heatwaves are now ten times more likely than they were before the year 2000.

Heatwave An extended period of hot weather relative to the expected conditions at that time of year

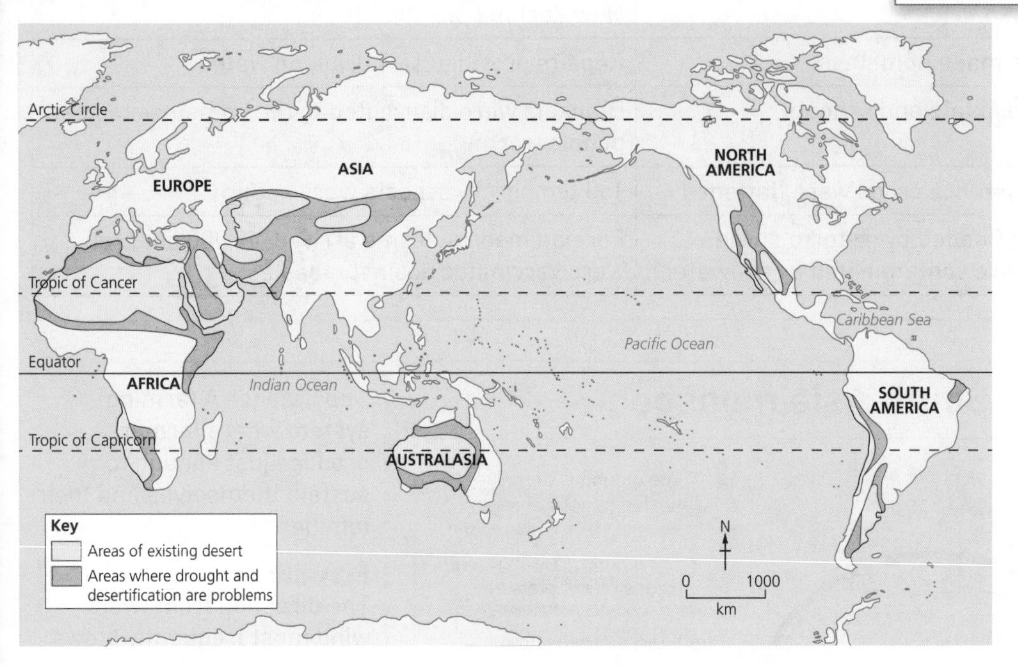

Figure 14 Location of key areas of drought since 2000.

A **drought** is a lack of precipitation in an area for a long period of time, often several months or even years. Droughts can happen anywhere, although in recent decades the most severe droughts have been located in:

- Australia
- South America – Brazil
- Africa – the Sahel (south of the Sahara Desert)
- Asia – areas of China and India
- The Mediterranean.

Climate change is predicted to lead to more frequent and severe droughts.

Drought A lack of precipitation in an area for a long period of time

How high-pressure systems lead to weather hazards

High pressure leads to light winds, dry weather conditions and sometimes drought. Droughts develop slowly, over a large area, with a beginning and end often difficult to identify. They are the result of:

- a lack of rainfall
- an environment, soil or bedrock which is poor at storing and retaining water
- hot weather which increases evaporation of water.

Exam practice

Explain why tropical low-pressure systems are described as the most destructive of weather hazards. [6]

Exam tip

Identify key terms in a question and plan your answer to address each one. In this question the key terms are 'tropical low-pressure systems', 'weather hazard', 'destructive'. You could split your answer into three paragraphs, each addressing one key term.

Example of a high-pressure hazard: drought in the California

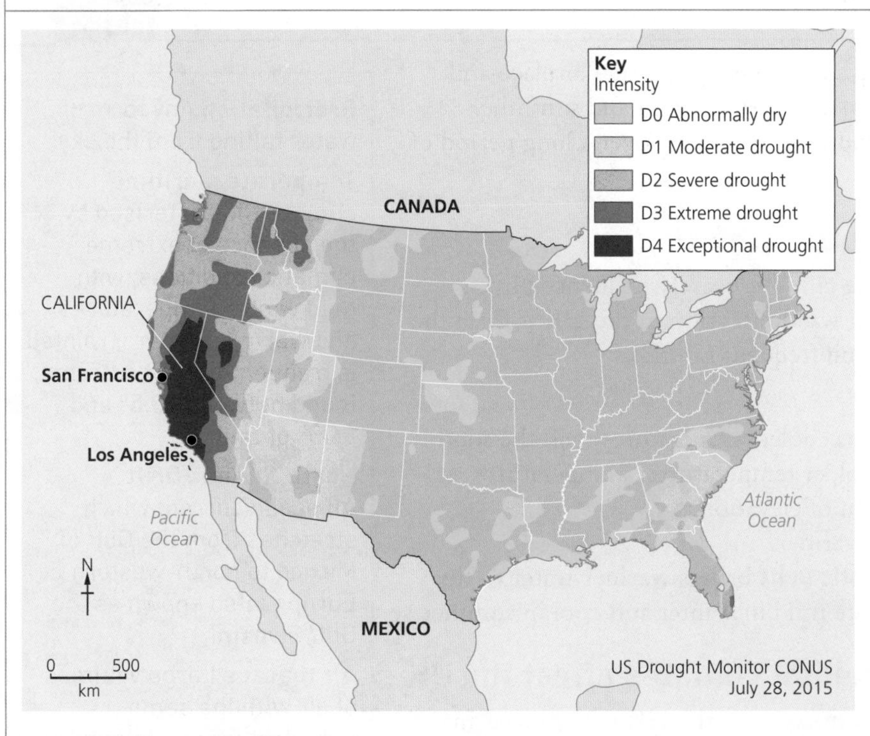

Figure 15 The extent of the drought across the USA in July 2015.

A drought lasting three years affected California between 2012 and 2015. An area of high pressure blocked low-pressure areas from bringing winter rainfall.

Social, environmental and economic impacts	Responses
Low river levels meant most hydroelectric power stations stopped producing electricity	Compulsory water restrictions including a ban on watering gardens and washing cars
California usually grows half the fruit and vegetables produced in the USA. Crop failure led to shortages and price rises	Toilets, washing machines and showers must use modern, low-water technologies
17,000 agricultural jobs were lost	A reduction in the amount of electricity generated through hydroelectric power
Salmon and trout died as river levels fell and water temperatures rose	Investment into desalination plants that remove salt from seawater
Groundwater supplies were not recharged. This could have serious long-term impacts on farming which relies on this water to irrigate crops	Farmers began growing crops that do not need so much water

Revision activity

Build up a Five Ws case study of the California drought. The Five Ws are questions whose answers are considered basic in information gathering or problem-solving. They constitute a formula for getting the complete story on a subject.

- **What** happened?
- **When** did it take place?
- **Where** did it take place?
- **Why** did it happen?
- **Who** did it affect?

Each question should have a factual answer. Use the information you have gathered to complete your own newspaper-style report on the California drought.

Now test yourself

1 Locate the Pacific island of Vanuatu and the US state of California on a blank map of the world.
2 Give three impacts of the low-pressure hazard in Vanuatu and three impacts of the high-pressure hazard in California.

TESTED

What factors create variations in weather and climate within the UK?

REVISED

Weather describes the atmospheric conditions at a particular place and time. These include temperature, **precipitation**, wind and sunshine. Climate is the average weather conditions measured over a long period of time, at least 30 years.

What factors affect climate in the UK?

The UK has a **temperate maritime climate** heavily influenced by latitude and the sea. The prevailing wind direction in the UK is from the south-west. This brings moisture and frequent rainfall.

Factors affecting climate include:

- Latitude: the north of the UK has cooler temperatures than the south.
- Altitude: mountain areas have colder temperatures. Temperatures decrease by 1 °C for every 200 m of elevation.
- Aspect: south-facing slopes are warmer.
- Ocean currents: the **North Atlantic Drift** brings warmer water to the UK's shores and keeps the climate mild in winter and cool in summer.

How maritime and continental climates affect the UK

Wind direction brings different **air masses** to the UK which have an important effect on the weather:

- north-westerly brings polar maritime air: cool and showery
- south-westerly brings tropical maritime air: mild and wet
- south-easterly brings tropical continental air: hot and dry
- easterly brings polar continental air: hot in summer and cold in winter
- northerly brings Arctic air: cold with snow in winter.

The west of the UK experiences a more maritime climate than the east, particularly in the winter months.

Weather associated with low-pressure

- Low-pressure systems, or depressions, produce cloudy, rainy and windy weather, and often affect the UK, particularly in winter months.
- Depressions begin in the Atlantic, moving eastwards across the UK.
- Within the low pressure there are usually **fronts** where one air mass rises above another.
- As air rises and cools, water vapour condenses to form clouds and precipitation.

Weather associated with high-pressure

- High-pressure areas or anticyclones give low wind speeds and stable conditions with no clouds.
- In summer, anticyclones usually bring dry and hot weather, which may lead to drought. Rapidly rising warm air may result in **convectional rainfall** and thunderstorms.
- In winter, clear skies bring cold nights and perhaps ice and fog.

Precipitation Any form of water falling from the sky

Temperate maritime climate Characterised by the absence of extreme climatic conditions, with mild winter temperatures and warm summers; rainfall is frequent but not extreme; found between 23.5° and 66.5° of latitude

North Atlantic Drift An ocean current which stretches from the Gulf of Mexico to north-western Europe (also known as the Gulf Stream)

Air masses Large volumes of air with the same temperature and humidity throughout

Front The boundary where two air masses meet

Convectional rainfall When the land warms up and heats the air above it, causing the air to expand and rise. As the air rises it cools and condenses, and large cumulonimbus clouds are formed, leading to heavy rainfall. This type of rainfall is common in tropical areas

Urban microclimate

- Urban heat islands: the result of the release and reflection of heat from buildings and the absorption by concrete, brick and tarmac of heat during the day, and its release at night.
- Urban precipitation: in the summer months, the extra heat generated may cause air to rise, which leads to convectional rainstorms.
- Urban winds: urban areas are less windy than surrounding rural areas.

Factors influencing microclimate

Microclimate is the climate of a small-scale area, such as a garden, valley or city. In the UK distinctive microclimates include:

- uplands
- coasts
- forests
- urban areas.

Revision activity

Complete the table below to explain frontal rainfall:

Number	Explanation
1	Warm air is lighter than cold and it rises above the colder air.
2	
3	

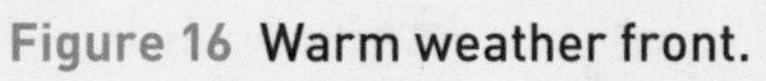

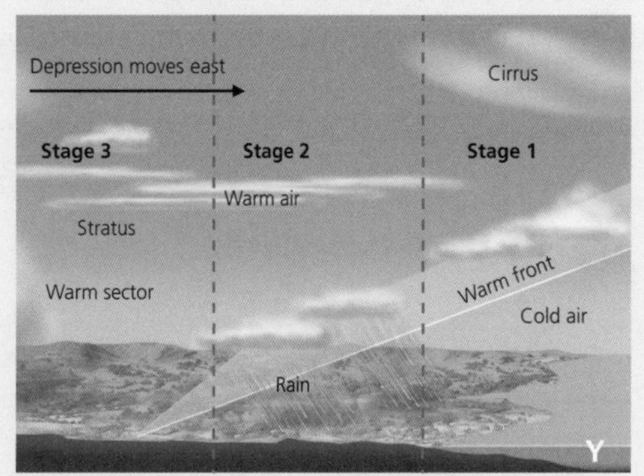

Figure 16 Warm weather front.

Now test yourself

TESTED

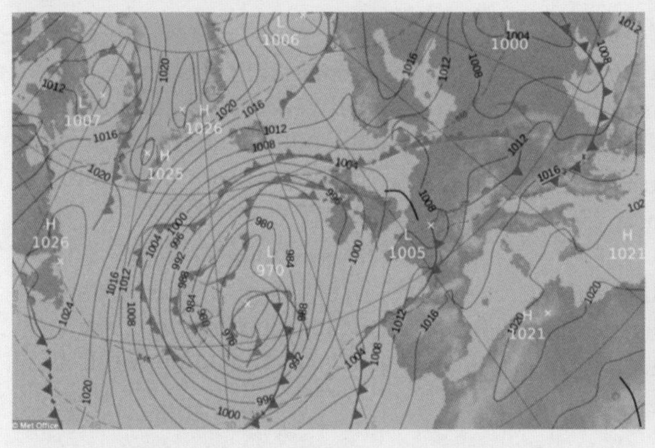

Figure 17 Weather on 20 October 2014.

Use evidence from the satellite photograph and weather map to write a weather forecast for the UK for 20 October 2014.

Exam practice

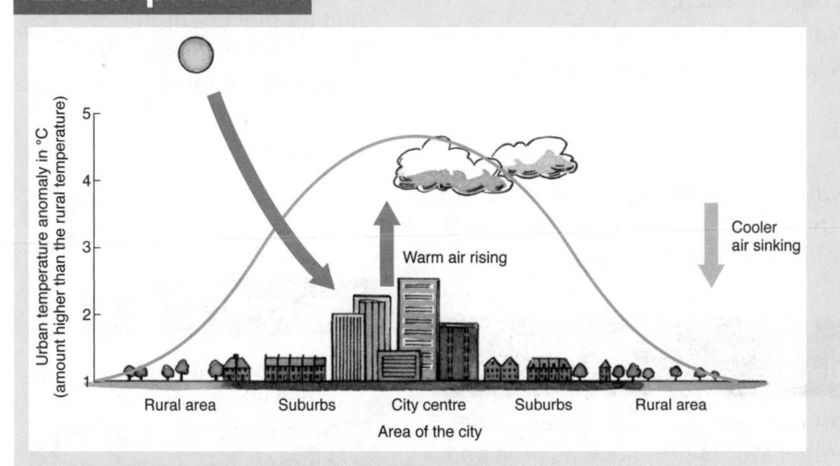

Figure 18 Why it rains more often in cities.

Annotate the diagram to explain how temperatures and rainfall totals differ in urban areas compared to rural areas. [4]

Exam tip

Labelling is when you name something, for example 'warm air rising'. Annotating is where you further explain your point, for example 'warm air rising leads to an increase in convectional rainfall'.

Processes and interactions within ecosystems

Where are large-scale ecosystems found?

REVISED

Ecosystems can exist at a variety of scales from a rock pool at the seaside to global systems such as the tropical rainforest.

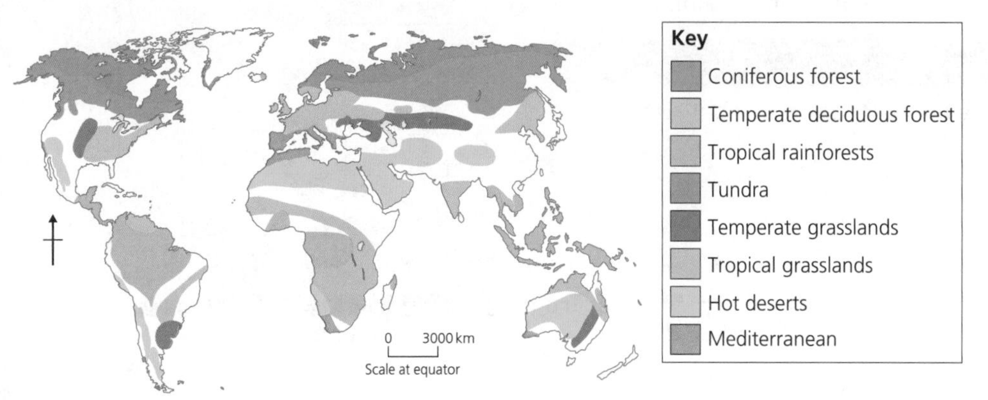

Figure 19 Global distribution of the major biomes.

Characteristics of large-scale ecosystems

Large-scale **ecosystems** or **biomes** develop over a very long period of time. Climate is the most important factor in determining their distribution:

- Rainfall: the amount and seasonal patterns are key to the distribution of all biomes, for example a place that receives less than 25 cm of rain per year is described as a desert.
- Temperature: when rainfall is reliable and distributed evenly throughout the year then temperature becomes the most important factor, for example mountains have a very distinctive ecosystem since temperatures fall by 1 °C for every 200 m of altitude.

Relief, geology and soils are other important factors in the distribution of biomes. Examples include the following:

- Tropical rainforest: located either side of the equator where hot and wet conditions encourage continuous growth of plants.
- Temperate deciduous forest: a hot and a cold season means that trees lose their leaves in the autumn to conserve energy.
- Coniferous forest (taiga): it is so cold in winter that trees have evolved needle-like leaves to survive the frosts and reduce moisture loss.

Ecosystem The links between plants, animals and the non-living things around them such as rocks, soil, water and climate

Biome A large-scale ecosystem where climate, vegetation and soils are broadly the same within the area

Revision activity

Draw a table on a sheet of A5 card with four columns, as below. Carry out further research and complete the table by giving important facts about each of the major biomes identified on the map above. One has been completed as an example:

Biome	Distribution	Climate	Plants and animals
Coniferous forest	Found between 50° and 60° north of the equator and in mountain areas	Long winters, temperatures well below freezing. Heavy snowfall	Coniferous trees such as pine and animals such as reindeer

Now test yourself

Imagine you are a TV producer making a programme on the plant and animal life of the UK. Write an introduction for the presenter of the programme about the temperate deciduous forest ecosystem.

TESTED

Exam practice

1 Use the map above to describe the distribution of the tropical rainforest biome. [4]

2 Give two features of the climate of the tropical rainforest ecosystem. [4]

What are the key processes of ecosystems?

REVISED

Processes and relationships that link living and non-living parts of ecosystems

The living (biotic) and non-living (abiotic) parts of the ecosystem are interdependent. If one part of the ecosystem is altered then the whole ecosystem will be changed. Humans are increasingly having an impact on natural ecosystems.

Succession

Ecosystems develop over time through succession:

- Hardy pioneer species colonise a bare area of ground.
- Weathered rocks and the decay of plants increase the supply of nutrients and allow new plants to grow which support insect and animal life.
- Soils become deeper which allows bigger plants and a greater variety of plants to live in the area.
- Given enough time, a dominant species, such as oak trees, will invade and the succession is complete. The ecosystem has achieved a sense of balance and although there are many daily changes the overall mixture of plants and animals species remains stable. This is the climax community.

Important processes in an ecosystem include the carbon cycle (see page 76), **water cycle**, **nutrient cycle** and **food web**. The greater the inputs of water, nutrients and energy, the greater the volume and diversity of plants and animals that can be supported.

Water cycle The movement of water between the stores of water in the hydrosphere, lithosphere, atmosphere and biosphere

Nutrient cycle The movement of nutrients in the ecosystem between the stores in the hydrosphere, lithosphere, atmosphere and biosphere

Food web The system of interlocking and interdependent food chains

Leaching The process of washing out from soils of soluble nutrients

Food chain The interconnections between different organisms (plants and animals) that rely on one another as their source of food

The water cycle

The water cycle is the journey water takes as it moves from the land to the sky and back again. It follows a cycle of evaporation, condensation and precipitation. On this journey it is absorbed by plants and consumed by animals.

The nutrient cycle

Nutrients are stored in water, rocks and the atmosphere and they flow through an ecosystem in a cycle:

- weathered rock releases nutrients into the soil
- water is added to the soil by rainfall
- plants absorb the nutrients through their roots and leaves
- herbivores gain nutrients by eating plants
- plants and animals die and are decomposed by bacteria and fungi
- nutrients are returned to the soil.

Nutrients may be lost to the system by **leaching** and by being washed away in surface runoff.

Food webs

- The Sun is the source of all energy for all life on Earth. It provides both heat and light energy.
- Plants (producers) convert energy from the Sun into food through photosynthesis.
- Herbivores (primary consumers) eat the plants.
- Carnivores (secondary consumers) eat the herbivores and so energy moves through the **food chain**.
- Food chains are connected to make a food web.
- The number of living organisms decreases at each stage of the food chain because energy is lost, used up in transpiration, movement and breathing.

Example of a biome: the rainforest ecosystem

Tropical rainforests are found along the equator in areas such as Brazil, Congo and South-East Asia.

- They cover about six per cent of the total land area.
- They contain the most diverse range and highest volume of plant and animal life found anywhere on Earth.
- There is a continuous growing season. Rainfall totals often exceed 2000 mm a year and temperatures are high, averaging 25 °C.
- Vegetation is divided into layers: emergents, canopy, undercanopy, shrubs and the ground layer.

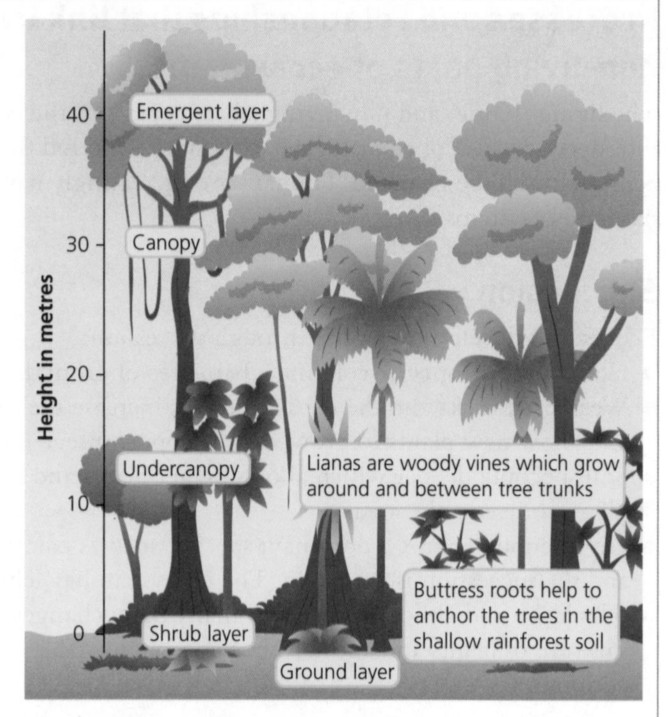

Figure 20 **Structure of the rainforest.**

Rainforest nutrient cycle

- The nutrient cycle in the forest is very delicate: about 80 per cent of the rainforest's nutrients come from trees and plants, only twenty per cent of the nutrients are stored in the soil.
- Hot, damp conditions allow for rapid decomposition from leaves that fall. This provides plentiful nutrients that are easily absorbed by plant roots.
- These nutrients are in high demand from the rainforest's many fast-growing plants. They do not remain in the soil for long and stay close to the surface of the soil.
- Consumers (herbivores) eat plants. When they die nutrients are returned to the soil by decomposers. Nutrients are quickly recycled.

Rainforest water cycle

- Roots of plants take up water from the ground and rain is intercepted as it falls, much of it at the canopy level.
- As the rainforest heats up, the water evaporates into the atmosphere and together with transpiration forms clouds to make the next day's convectional rainfall.
- The forest is a great store of water between the rainfall events.
- Rainfall feeds some of the great rivers that leave these forests such as the Amazon River.

Rainforest carbon cycle

- Plants during photosynthesis absorb CO_2 from the surrounding atmosphere.
- Rainforest leaf systems comprise about 70 per cent of the world's total leaf surface area and account for between 30 and 50 per cent of the Earth's total primary productivity. Rainforests store more carbon per unit area than any other type of ecosystem.
- When a plant respires or dies the carbon within the plant is released back into the environment. Some plants are eaten by herbivores who in turn may be eaten by carnivores. Carbon is then released by animal respiration and when the animal dies.
- The soil becomes an important store of carbon from the dead plants and animals.
- Tropical forests are critical in the carbon cycle of the planet. When forests are cleared and burned, 30–60 per cent of the carbon is lost to the atmosphere.

Revision activity

1. Complete a set of ten flashcards with questions and answers of ways in which plants and animals have adapted to life in the rainforest. Now do the same for life in the savannah grassland.
2. Use a double-bubble diagram such as the one on page 32 to compare the rainforest ecosystem with the savannah grassland ecosystem. Compare their location, climate and examples of plant and animal adaptation.

Example of a biome: the savannah grassland

Savannah grasslands are found within the tropics in a broad band about 5–15° north and south of the equator, between the tropical rainforests and the hot deserts of the sub-tropics, in countries such as Brazil, Tanzania, India and northern Australia.

- The climate has a marked wet and dry season. The wet season occurs in 'summer' with heavy convectional rain. Temperatures are high throughout the year, ranging between 23 and 28 °C.
- Vegetation consists of scattered trees, drought-resistant bushes and grasses that can grow up to 4 m in height.
- Drought and fire dictate what species survive; plants are therefore **xerophytic** and **pyrophytic**.
- The savannah contains a huge variety of plants, insects, birds and animals. This **biodiversity** makes the savannah lands a popular tourist destination.

Savannah nutrient cycle

- The store of nutrients in the biomass is less than that in the rainforest because of the shorter growing season and because fire returns carbon to the atmosphere.
- Nutrients are stored near the surface of soils since it comes from decayed organic matter (vegetation) from the previous growing season. This organic matter decays rapidly due to the high temperatures.
- The role of fire, whether natural or made by humans, is vital as it helps to maintain the savannah as a grass community. It mineralises the litter layer, kills off weeds and prevents trees from growing.
- The most important nutrient recyclers are termites that gather up vegetation and process it in one spot. The termite mounds become a nutrient hotspot.

Savannah water cycle

- The savannah experiences recurrent episodes of drought lasting 4–8 months of the year.
- During the xeropause, or 'dry spell', plant activities (growing, dying, decomposing) continue, but at vastly reduced rates.
- Resistance to drought is more important to savannah vegetation than resistance to fire.
- Diverting water for tourists is exploiting local water reserves, leaving plants and animals short of water. Tourist hotels sometimes dump waste into rivers.

Savannah carbon cycle

- Tropical savannahs cover around twenty per cent of the Earth's land surface and while they have fewer trees and stored carbon than rainforests or temperate forests, their extent makes them significant in the global carbon cycle.
- Regular bush fires release many tonnes of CO_2 into the atmosphere.

Now test yourself

TESTED

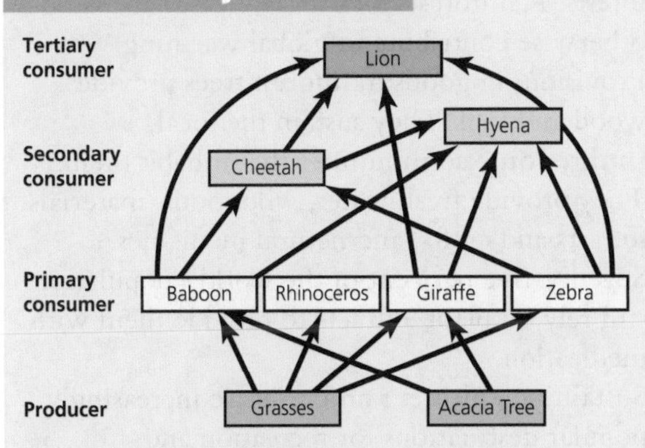

Figure 24 **Food web for the semi-arid grassland ecosystem in Africa.**

1 Give an example of a food chain in the savannah ecosystem.

2 Complete the table:

Living part of the ecosystem	Rainforest	Savannah
Tertiary consumer		
Secondary consumer		
Primary consumer		
Producer		

Xerophytes Plants that can survive drought

Pyrophytes Plants that have adapted to tolerate fire

Biodiversity A measure of the variety of plants and animals that live in an ecosystem

What is biodiversity?

Biodiversity is a measure of the variety of plants and animals that live in an ecosystem. Areas which have a rich biodiversity include rainforests, for example, while a hot, dry desert would have a low biodiversity.

Endemic species are those which are unique to a given region or location and are seen nowhere else in the world.

Those places on the Earth that have a rich biodiversity and a large number of endemic species are particularly important to conserve.

Endemic A species which is only found in a given region or location and nowhere else in the world

Key services The processes by which the environment produces resources used by humans such as clean air, water, food and materials

Indigenous Originating in a particular area, region or nation; usually applied to flora, fauna and people

Ways in which ecosystems provide key services

Many believe that all ecosystems should be protected, not just for their scientific value but because they provide people with essential or **key services**. These include:

- provisioning services: timber for building, food and clean water
- regulating services: flood control, preventing soil erosion and providing medicines
- cultural services: spiritual and leisure resources
- supporting services: carbon cycling that maintains the conditions for life on Earth.

Examples of key services provided by different ecosystems

- Coniferous forests (taiga): provide a huge source of timber.
- Savannah grasslands: provide food and building materials for nomadic people such as the Maasai. Wildlife attracts tourists and helps poorer countries to develop.
- Peat bogs: act as huge stores of CO_2 so helping to regulate the greenhouse effect.
- Sand dunes: act as natural coastal defences against storm surges, strong winds and coastal floods.

Now test yourself

Make a list of the key services provided by the tropical forest ecosystem both to the indigenous people who live in the forest and to the rest of the global population.

TESTED

Example: key services provided by rainforests

- Nutrient cycling: a rainforest recycles the nutrients it has created; without tree cover these nutrients would be lost. Plants anchor the soil and prevent it, as well as the nutrients within it, from being washed away by heavy rainfall.
- Rain making: some rainforest services extend across vast areas, for example the Amazon forest makes as much as 50 per cent of its own rainfall.
- Regulating climate and air quality: rainforests continually recycle huge quantities of water, feeding rivers and lakes. Millions of people live downstream of rivers that leave the forests. Rainforests lock away CO_2 that would otherwise contribute to global warming.
- Provisioning goods: rainforest trees provide wood and fuel. They sustain the local, **indigenous**, communities that inhabit them. They provide fresh water, wild foods, materials for arts and crafts, and natural medicines. Seventy-five per cent of the world's population still rely on plant extracts to provide them with medication.
- Sustaining culture: rainforests are increasingly popular destinations for recreation and ecotourism and they hold considerable educational and scientific value.

Example of a small-scale ecosystem: sand dunes

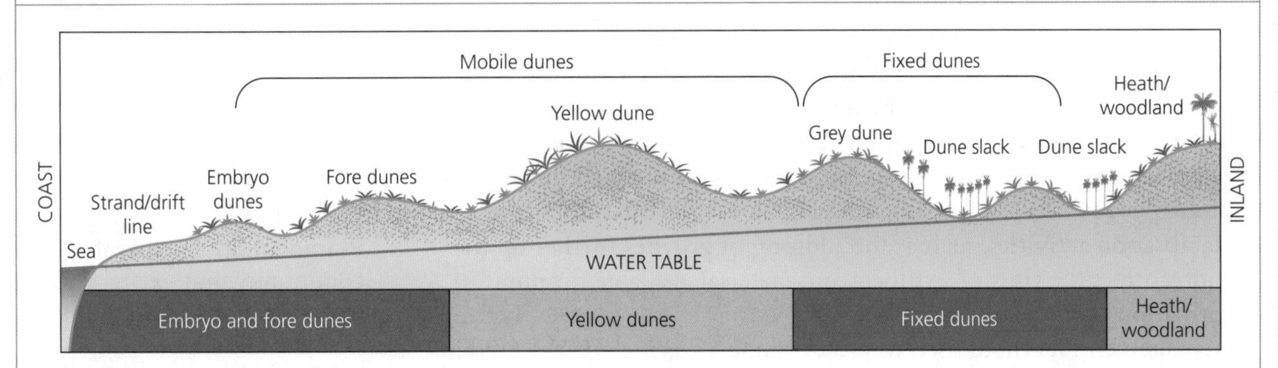

Figure 21 Sand dune ecosystem.

- Sand dunes are accumulations of sand stabilised by vegetation.
- Their development needs a source of sand and an onshore prevailing wind.
- Grains of sand stop when they hit an object such as seaweed or a lump of driftwood.
- A small dune, referred to as an **embryo dune**, characterised by pioneer plant species such as sand rocket, is formed.
- As the pioneer plants grow, the small dunes become stabilised by root systems. The dune gets bigger. Plants will die and decompose, adding nutrients and organic matter. New plants are able to grow in a process known as succession.
- Larger dunes called **fore dunes** develop (also known as mobile or yellow dunes). Fore dunes are dominated by species such as marram grass.
- In time, the dunes stop growing due to a lack of wind-blown sand. They are then referred to as **fixed or grey dunes** and a wide variety of plants such as orchid, heather and bramble cover the dune.
- Eventually a climax community is established, when the vegetation is in a state of equilibrium with the environment and there is no further influx of new species. Climax community is dominated by heathers and birch or oak woodland.

In the sand dune ecosystem key services include:

- providing the opportunity to develop tourism
- acting as a natural coastal defence against storm surges and coastal flooding
- providing sand for building purposes.

How human activity affects biodiversity, flows, cycles and processes

Sand dunes are rich wildlife habitats for a wide variety of specialised plants and animals. They are delicate and, as a result, very susceptible to human interference:

- Recreation: sand dunes are an attractive place for many leisure activities, for example walking, pony-trekking, trail biking, bird watching and picnicking. Marram grass is trampled and is easily killed. Bare sand is then exposed to the wind and without the roots of the marram securing it, sand is blown away to leave large semi-circular hollows known as blow-outs. Blow-outs can persist for a long time, the rich habitat is destroyed and biodiversity reduced. Songbirds such as linnets and skylarks nest on the ground and are easily disturbed by visitors.
- Economic: farmers use dune areas for grazing animals, particularly cattle. Cow dung enriches the soil and affects natural processes and nutrient cycles. Intensive farming introduces fertilisers. Plant and animal diversity is reduced through soil compaction and nutrient enrichment. Sand dunes are also in demand from landowners, who wish to develop the land for tourist developments, such as caravan

Embryo dune The youngest dune at the front of the dunes, nearest the sea

Fore dunes Older and slightly higher dunes just shorewards of embryo dunes

Fixed or grey dunes Found further inland where conditions for plant growth improve parks and golf courses, or for housing. Direct removal of sand from the beach or the dunes for building purposes reduces the sand available to create and expand existing dunes.

- Environmental: afforestation has been used in some areas to prevent the movement of mobile sand. This affects biodiversity, natural succession and processes in the dunes. The introduction of rabbits by people has also had a significant impact on many dune areas by preventing the encroachment of trees and bushes and succession to reach the natural climax vegetation. This is an example of a **plagioclimax**. Coastal stabilisation activities, such as the addition of groynes further up the shore, affect the flow of sediment and material available for dune formation.
- Management: methods used to protect dune areas include the erection of protective fencing, building boardwalks, planting marram grass, putting up signs and the setting up of sand traps. Fencing prevents people from walking over the more unstable dunes. Boardwalks encourage people to avoid walking over the dunes and stick to the pathways. Signs alert the public of the damage they cause by walking on the dune, encouraging them to stay away. Marram grass planting and the construction of sand traps minimise the effects of trampling.

Plagioclimax The plant community that exists when human actions prevent the climatic climax vegetation being reached

Revision activity

1. Make a copy of the Figure 21 (page 91). Annotate your diagram to explain the creation of the sand dune ecosystem.
2. Draw a spider diagram to summarise the threats to the sand dune ecosystem from human activities.

Exam tip

Make sure that you understand the meaning of key terms such as distribution. Distribution refers to the way something is spread out or arranged over a geographic area.

Now test yourself

TESTED

1. What is meant by a climax community?
2. Describe the impact of leisure activities on biodiversity in a sand dune ecosystem.

Exam practice

Figure 22 The savannah.

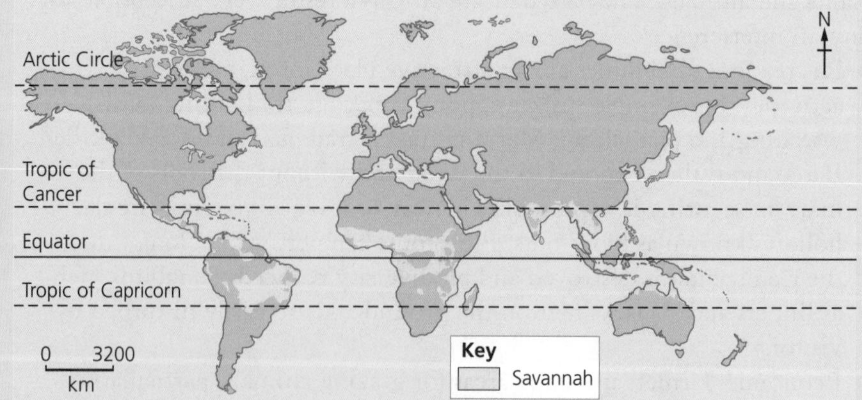

Figure 23 Location of the savannah ecosystem.

1. Describe the distribution of the savannah ecosystem. [4]
2. Give two features of the savannah ecosystem that makes it distinctive. [4]
3. Explain why climate is important in the development of the savannah. [6]

Human activity and ecosystem processes

How do people use ecosystems and environments?

REVISED

How people change ecosystems for food, energy and water

Human activity has had an impact on all ecosystems:

- In the UK, much of our temperate deciduous forest has been removed for farming and to provide room for cities.
- In the USA, much of the temperate grassland has been removed to grow cereals.
- In the Sahel region of Africa, Lake Chad has shrunk dramatically in the past 50 years due to the abstraction of water from rivers that supply the lake.
- The Mediterranean ecosystem and environment is under pressure from tourism and its demands on the limited water resources.
- The Arctic tundra ecosystem is under threat from drilling for oil in places such as Alaska.

Renewable Resources which either are never-ending or replenish quickly enough that their use does not lead to exhaustion

Non-renewable Resources considered to be finite as their rate of use far outstrips the rate at which they are formed

Example of an environment used for energy production: Gwynt y Môr offshore wind farm

Power can be generated from **renewable** or **non-renewable** sources. Non-renewable sources, such as oil and coal, are running out and scientists are exploring the potential of renewable sources, such as tides and the wind. Gwynt y Môr offshore wind farm, which opened in June 2015, is located 13 km off the coast of north Wales and 16 km from the seaside resort of Llandudno:

- it is the second largest operating offshore wind farm in the world
- it has 160 wind turbines, 150 m high
- it cost £2 billion.

Npower, one of the operators, states that the farm will produce enough power to supply 400,000 residential households per year.

The BBC reports that up to 100 jobs will be created at Mostyn docks, near Holywell, Flintshire, to service the turbines.

The Royal Society for the Protection of Birds (RSPB) states that windfarms can harm birds through disturbance and collision. Thousands of migrating birds may fly through the Gwynt y Môr windfarm area.

WWF Cymru supports the decision: 'We have a real opportunity with Gwynt-y-Môr to use the powerful resource off the Conway coast to help global efforts to fight climate change.'

The National Trust raised concerns because it owns three-quarters of the Welsh coastline. It fears for the future. More wind farms would spoil the beauty of the coastline it aims to preserve.

Many local people in Llandudno are opposed to the project because they describe the town as a historic place with natural scenic beauty. They see the project as an eyesore that will deter tourists.

Revision activity

Do you think the Gwynt y Môr wind farm should have been built? Prepare a speech that could be debated in the Welsh Assembly Government.

Now test yourself

Give an example of one way in which human activity has modified an ecosystem for food production.

TESTED

How do human activities modify processes and interactions with ecosystems?

REVISED

Example of impacts of human activity on a rainforest ecosystem: the Amazon rainforest

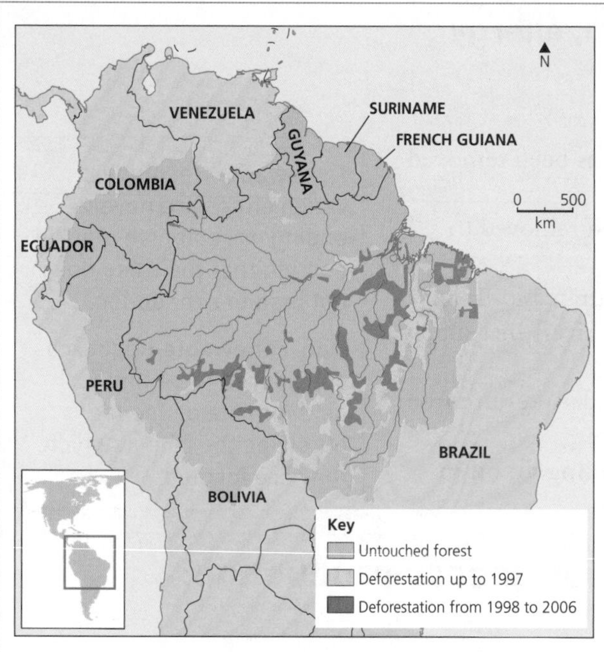

Figure 24 Deforestation in the Amazon basin.

The Amazon rainforest is 5.5 million km^2 in area and is located in South America. Fifty per cent of tropical rainforest has been destroyed in the past 100 years.

Reasons for destruction:

- Farming, for example commercial cattle ranching and growing crops such as soyabeans, oil palms and bananas (**monoculture**).
- Slash and burn farming, for example subsistence farmers clear the forest by burning. Nutrients in the soil are soon depleted. The land is abandoned and another patch of the forest cleared.
- Mining, for example in the Amazon there is the largest iron-ore mine in the world in Carajás.
- Roads and railways, for example a 900 km rail line links Carajás to São Luís on the coast to export iron.
- Logging, for example expensive woods, such as mahogany, are exported across the world.
- Electricity supplies, for example hydroelectric schemes require the construction of huge dams such as the Tucuruí Dam in Brazil.

Impacts – gains:

- Profits from developments such as Carajás have helped Brazil to become richer.
- Long-term jobs have been created in logging, farming, mining and tourism.
- The **multiplier effect** has encouraged other industries to locate in the area.
- Resettlement provides a better way of life for people who lived in shanty towns and relieves population pressure in the cities.

Impacts – losses:

- Native Indians such as the Yanamami have lost their homelands, culture and way of life.
- Animals and plants have lost their **habitats**, leading to extinction. Some of these species are used for medicinal purposes, for example the rosy periwinkle is used for leukaemia treatment.
- Changes in climate: interception and transpiration are reduced and evaporation is increased, leading to a drier regional climate. Water and silt pour into rivers, causing them to flood.
- When trees are chopped down, the nutrient cycle is broken. Rain removes nutrients from the soil via surface runoff and leaching. Soils quickly become infertile. Soil dries up in the sun. When it rains, it washes the soil away and there is gullying and mud slides on steep slopes. The rainforest never fully recovers, and wildlife and plant life are reduced.

Revision activity

Complete a summary on a sheet of A5 card broken down into these headings:

- title
- location of the example (including a map like the one above)
- at least three reasons for the destruction of the rainforest
- at least six ways in which humans have 'gained' or 'lost' because of this destruction.

Example of impacts of human activity on a savannah grassland ecosystem

In the savannah grasslands, indigenous people are often nomadic and keep herds of sheep, cattle or goats. Some live in permanent villages and grow crops. Population growth is rapid in many areas. The population of the **Sahel** region is growing at three per cent a year and doubling every twenty years.

A huge problem in savannah grasslands is **desertification**. This may be partly due to climate change and reduced rainfall but it is also the result of overgrazing and overcultivation. Soil is losing its fertility and is either blown away in the dry season or washed away in the heavy rains in the wet season.

Trees are scattered in the savannah but they protect the soil:

- Trees recycle water into the atmosphere through evaporation and transpiration.
- Slash and burn agriculture removes trees, reduces evapotranspiration, interrupts the water cycle and reduces rainfall totals.
- The nutrient cycle is broken since leaf litter can no longer fall on the soil. Soils lose their fertility and structure since they lack an input of organic matter.
- The soil is eroded because there is no interception and because root systems have been removed. Heavy rainfall flows over the surface and picks up soil particles. It uses these to erode a deep channel in the land. This is known as **gully erosion**.
- In slash and burn agriculture, crops are grown for between one and three years, the land is abandoned and the soil regains its goodness during a **fallow** period. Overcultivation stops the soil recovering its organic content; it becomes dusty and eroded by wind and rainfall.

Habitats are being destroyed and the biodiversity of savannah grasslands is being reduced. Many tourists also visit the African savannahs on wildlife safaris, for example in Kenya. The wildlife is often disturbed and the people and vehicles damage the landscape.

Monoculture The cultivation of a single crop in a given area

Multiplier effect The 'snowballing' of economic activity, for example new jobs are created, people who take them have money to spend in shops, which means that more shop workers are needed

Habitat The natural home or environment of an animal, plant or other organism

Sahel A region in north-central Africa south of the Sahara Desert in an area prone to drought

Desertification The spread of desert, or desert conditions, from an established desert area into the surrounding area

Gully erosion Channels formed on a poorly vegetated hillside by soil erosion

Fallow A field left to naturally regain its nutrients after growing crops for a number of years

Exam practice

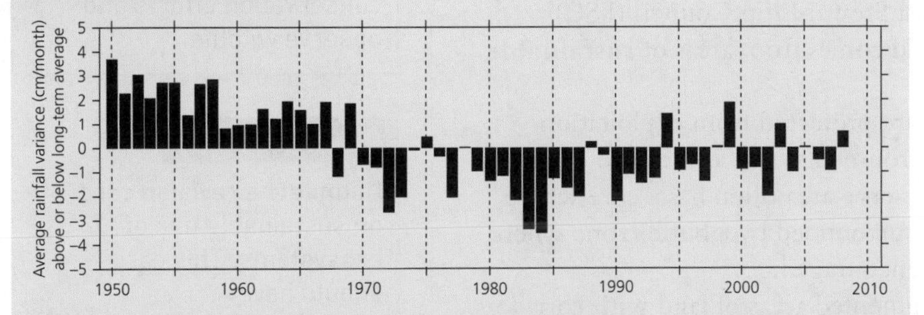

Figure 25 Changing rainfall in the Sahel region.

Describe how the patterns of rainfall have changed in the Sahel region between 1950 and 2010. [4]

Now test yourself

1. Explain what is meant by the multiplier effect.
2. Explain why the nutrient cycle in the rainforest is delicate.
3. Explain why overgrazing and overcultivation are leading to desertification in the savannah grasslands.

TESTED

Theme 5 Weather, Climate and Ecosystems

How changes may impact on biodiversity

Habitat loss due to destruction, fragmentation or degradation is the primary threat to biodiversity. When an ecosystem has been dramatically changed by human activities, such as agriculture, oil exploration or water diversion, it may no longer be able to provide food, water, cover and places to raise young. Every day there are fewer places left that wildlife can call home.

There are three major kinds of habitat loss:

- Habitat destruction, for example deforestation, draining wetlands and damming rivers.
- Habitat fragmentation, for example by urban development, dams and water diversions. Habitats become divided into separate fragments, creating isolated populations. These **ecological islands** affect biodiversity by reducing the amount of suitable habitat and by restricting migration between different communities and hence opportunities to breed and feed.
- Habitat degradation, for example pollution, introduction of new species and disruption of ecosystem processes such as nutrient cycles.

Revision activity

Find out what is meant by slash and burn agriculture.

How can ecosystems be managed sustainably?

REVISED

Ecosystems provide key services that include:

- clean water in rivers
- reducing the risk of flooding
- providing natural resources such as timber and air to breathe
- producing foodstuffs such as fish and honey.

Conservationists would argue that maintaining these key services is more valuable in the long term than the short-term gains from the unsustainable exploitation of ecosystems.

Example of sustainable use of the rainforest

- Agro-forestry: growing trees and crops at the same time.
- Selective logging: only cutting down trees that are a certain species or age. The international Forest Stewardship Council (FSC) gives assurances that timber sold comes from areas of **sustainable** logging.
- Forest reserves: areas of forest are protected from exploitation and maintained as natural environments such as the Alto Maués National Park in Brazil. The reserve area often has a core where no human activity takes place surrounded by a buffer zone where sustainable use of the forest is encouraged.
- Wildlife corridors: link up fragmented areas of land with corridors of vegetation so that animals can find food and can find a mate to reproduce.
- Encouraging **ecotourism**, a rapidly growing form of tourism and highly profitable.
- Debt-for-nature swaps: poor countries can 'swap' the debt they have with a rich country if they protect an area of rainforest.

Ecological island An area of land, isolated by natural or artificial means from the surrounding land

Sustainable Using ecosystems to meet the needs of the current generation without compromising the needs of future generations

Ecotourism Tourism directed towards exotic natural environments, intended to support conservation efforts and observe wildlife

Revision activity

Complete a revision card on sustainable use of ecosystems. This card should have:

- a definition of the term 'sustainable'
- at least four ways in which the rainforest can be sustainably managed and four ways in which savannah grasslands can be sustainably managed.

Example of sustainable use of the savannah grassland

Solutions suggested to solve the problem of soil degradation and desertification include:

- Crop rotation and fallow periods: grow different crops each year so that different nutrients are used, and allow fallow periods so that soil can regain its fertility.
- Shelter belts: plant shelter belts, areas of forest or hedges, to protect farmland from the effects of water and wind erosion.
- Reforestation and afforestation: reducing wind and water erosion.
- Irrigation: water areas of land that have become arid by diverting rivers or digging wells, although if water is not used sustainably then this can cause water shortages elsewhere.
- Reduce grazing numbers: place limits on the number and types of animals that can graze on land, reducing the destruction of vegetation and eventual desertification.
- Population control: if population growth can be controlled then less agricultural land is needed and the intensity of farming will be reduced.
- Drought-resistant crops: **genetically modified** crops can be engineered to withstand poor soil and water shortages, such as new varieties of maize.

Example of the Great Green Wall

In 2010, eleven countries signed an agreement to plant a Great Green Wall, 15 km wide and 8000 km long, across the width of Africa, south of the Sahara. It aims are:

- To encourage local communities to plant native trees to create a living wall of trees and bushes. Trees would include nut and fruit trees and provide wild foods. Small fields between the trees can be planted with food and cash crops – a type of farming known as **agro-forestry**.
- To prevent desertification and soil erosion.
- To protect water sources, such as Lake Chad.
- Restoration or creation of habitats for plants and animals.

Niger and Senegal have made good progress in the building of the wall. Benefits have included:

- increased crop yields
- better fed livestock
- trees provide medicines and firewood.

In the other nine countries progress has been slow.

Genetic modification The placing of a gene from one organism into another so that the latter can take on a quality of the former that it doesn't otherwise have

Agro-forestry A land-use management system in which trees or shrubs are grown around or among crops or pastureland

Exam practice

Give two ways in which the rain forest can be used sustainably. [4]

Exam tip

Questions which are worth 6 marks are marked with a banded mark scheme, similar in style to the one below:

Band	Mark	Descriptor
3	5–6	Detailed and specific description of ways in which climate is important in the development of the savannah ecosystem
2	3–4	Description of ways in which climate is important in the development of an ecosystem. Maximum band 2 if no named examples
1	1–2	Valid but simplistic description or list of points
	0	Answer is incorrect or irrelevant

It is important in such questions to plan an answer, focus on the key words in a question and ensure that specific examples are used to gain a band 3 mark.

Now test yourself

1. What is the Great Green Wall?
2. Look over Theme 5. Write down key terms on one set of sticky notes and the definitions on a second set. Mix up the pads and then try to match each keyword to its definition.

TESTED

Theme 6 Development and Resource Issues

Measuring global inequalities

What are the global patterns of development?

REVISED

What is development?

Development is the process of change which improves the wealth and quality of life of people:

- Economic development is the result of increased employment and rising incomes.
- Social development occurs when there is rising life expectancy and greater access to education, healthcare, clean water and housing for all people, particularly for women and minority groups.
- Political development involves forming a stable government and allowing greater freedom of speech.

Development gap

The **development gap** that exists between richer and poorer countries was first described by Willy Brandt, a German politician. The Brandt Report was published in 1980 and divided the world into the less developed 'global south' and the more developed 'global north'. The line dividing the richer and poorer countries became known as the 'north–south divide' or the Brandt line.

Development gap
The widening difference in levels of development between the world's richest and poorest countries

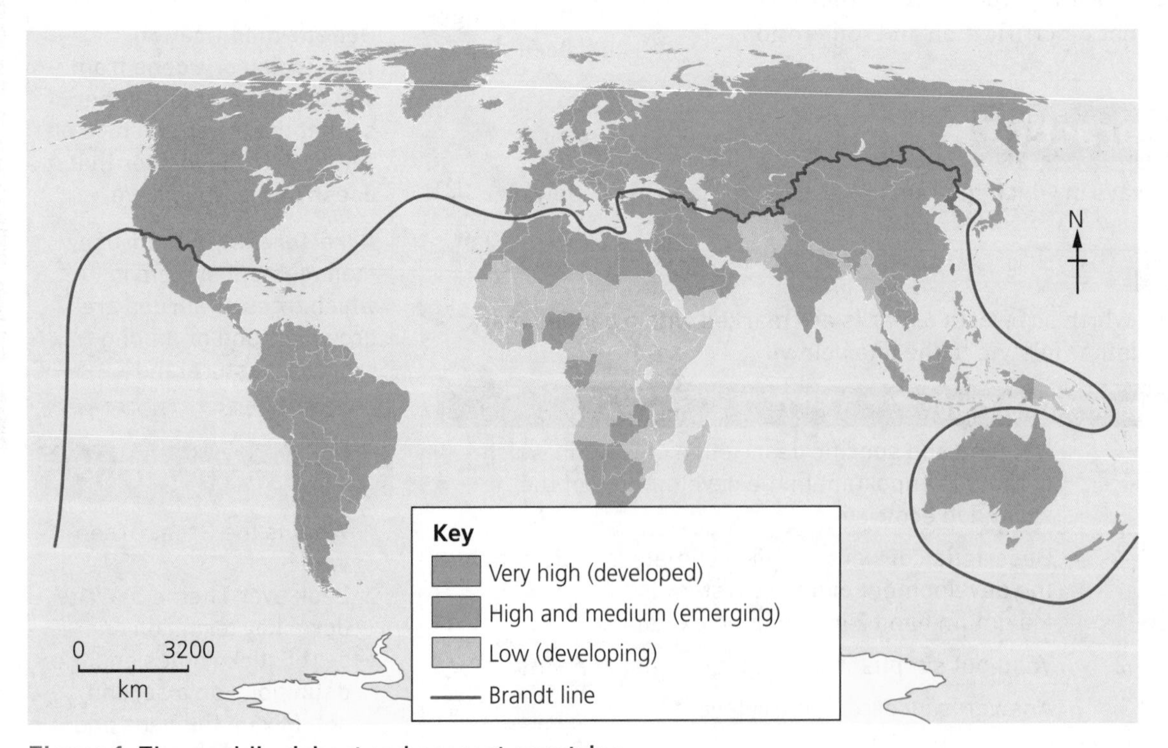

Figure 1 The world's richest and poorest countries.

Development continuum

It is simplistic to divide the world into rich and poor. In reality there is a **development continuum**, a sliding scale, from the very poor countries to the super-rich countries. Countries can move up and down this continuum. The **World Bank** divides countries into four categories of wealth according to their **gross national income (GNI) per capita**:

- High-income countries (HICs) with a GNI of $12,736 or more.
- Upper-middle-income countries with a GNI between $4126 and $12,735.
- Lower-middle-income countries with a GNI between $1046 and $4125.
- Low-income countries (LICs) with a GNI of $1045 or less.

Development continuum A linear scale from 'highly developed' countries to those with a 'low level' of development

World Bank An international financial institution that provides loans to developing countries for capital programmes

Gross national income (GNI) per capita A measure of the total economic output of a country, including income from foreign investments, divided by its population (per capita means per person)

Global patterns of development

In 2016, Oxfam published a report into global inequality in which it highlights the development gap between rich and poor people, not countries. It describes how in poor countries there are rich people and in rich countries there are poor people. Oxfam also noted that 62 people in the world own the same amount of wealth as the poorest 3.5 billion.

Many would argue that wealth alone does not give an accurate picture of human development. Bhutan, a country in South Asia, measures development through happiness, measuring the gross national happiness (GNH) of its population.

Revision activity

1 Which continent has the greatest number of low-income countries?
2 Describe how patterns of development have changed since the Brandt line was drawn in 1980.

Now test yourself TESTED

Give two limitations of using gross national income per capita as a measure of development.

How is economic development measured? REVISED

In order to compare levels of development in different places geographers use a number of indicators. Economic development indicators include:

- **Gross domestic product (GDP)**: the total value of all goods and services produced within a country in a year.
- Gross national income (GNI) per capita: the average wage of a country's population.
- Employment structure: the type of work people do, for example farming, manufacturing or services.
- Poverty: the percentage of the population who earn less than $1.90 a day (World Bank's 2015 measure).

Gross domestic product (GDP) The total value of goods and services produced by a country in a year

Limitations of evidence of development

Limitations of using these methods of measuring development include:

- They measure only wealth and do not include social factors.
- They do not recognise inequality within a country.
- They do not consider the cost of living within a country and hence the amount of goods that can be bought with a given amount of money.

The United Nations uses the human development index (HDI) to measure development. This statistic combines figures for life expectancy, education and per capita income. HDI was created to emphasise that quality of life should be the ultimate measure of development of a country, not economic growth alone.

Revision activity

1 Suggest reasons to explain why is it important to be able to measure development.
2 Explain why the United Nations uses the human development index (HDI) to measure development.

Example of a low-income country (LIC): Malawi

Malawi in south central Africa has a population of 16.8 million:

- The local market for goods is limited since 90 per cent of the population earns less than $2 a day.
- People tend to live in remote rural areas and there are few roads and means of transport for goods.
- Plots of land are small, soils are infertile, weather is unfavourable and there is limited use of fertilisers. Productivity has not improved since the 1970s.
- Families are large and 2.8 million people suffer from chronic malnutrition.
- Education standards are poor, for example almost 30 per cent of children do not attend primary school even though it is free.
- The AIDS epidemic has affected twelve per cent of the population. Illness or injury is common.
- Malawi is a landlocked country and hence finds it difficult to trade with the rest of the world.
- The economy depends on the export of low-value primary goods such as tobacco, tea, cotton and sugar. Imports are dominated by high-value products such as oil and machinery.

Now test yourself

TESTED

'The development gap provides a static view of the difference in development between countries.' What is meant by this statement? Explain why the development gap is dynamic.

Revision activity

1. Give three reasons to explain why Malawi remains a low-income country.
2. Read the example box on page 102 on Malawi. Complete a PowerPoint presentation of six slides that you would use in a speech to an international conference in which you explain how Malawi's trade hinders its economic progress.

Exam practice

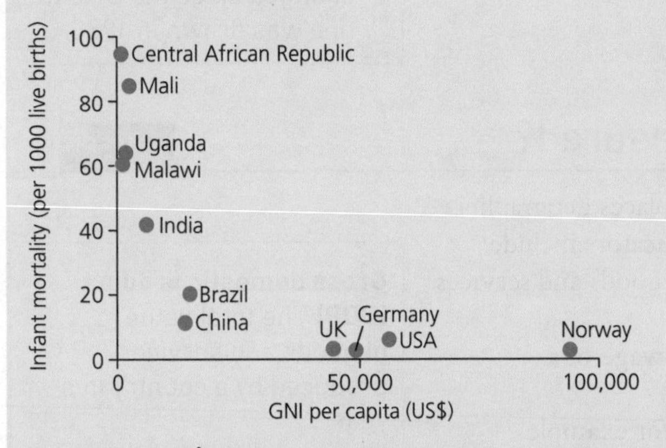

Figure 2 GNI/infant mortality rate per country.

1. Describe the relationship between infant mortality and GNI per capita. [3]
2. Suggest two reasons to explain this relationship. [4]

Exam tip

A scattergraph is used to test for a relationship between two variables. A line of best fit is drawn through the middle of the points to show the nature of the connection:

- If the points cluster in a band running from lower left to upper right, there is a positive **correlation**.
- If the points cluster in a band from upper left to lower right, there is a negative correlation.
- If it is hard to see where you would draw a line, and if the points show no significant clustering, there is probably no correlation.

Correlation The degree of association between two sets of data: can be either positive (as one increases so does the other) or negative (as one increases the other decreases)

Causes and consequences of uneven development at the global scale

What are the causes and consequences of uneven development?

REVISED

How has global trade led to uneven patterns of development?

Trade involves the buying and selling of goods and services between countries:

- HICs generally **export** valuable manufactured goods and services such as electronics, cars and financial services. They **import** cheaper primary products such as sugar, flowers, tea and coffee.
- In LICs the opposite is true. This means that LICs earn little and remain in poverty, the country is forced to borrow money to pay for its imports and the country goes into debt.

The price of primary products fluctuates on the world market. Prices are set in HICs and producers in LICs lose out when the price drops. LICs are dependent on the world trade system yet they have little control over how it operates.

Increasing trade and reducing their balance of trade deficit are essential for the development of LICs. Sometimes HICs impose **tariffs** (import duties) and **quotas** on imported goods and give **subsidies** to their own farmers.

Exports Goods and services produced in one country and shipped to another

Imports Goods and services brought into one country from another

Tariffs Taxes imposed on imports

Quotas Limits on the amount of goods imported

Subsidies Benefits given by the government usually in the form of a cash payment or a tax reduction

Interdependent Where countries are linked together in a complex web, economically, socially, culturally and politically, so that they are dependent on each other

Imports, exports and trade blocs

Free trade allows countries to import and export goods without tariffs or other barriers. It results in increased exports, more job opportunities, a greater choice of goods and lower prices for consumers.

Trading blocs, such as the European Union (EU), are groups of countries that work together to promote free trade between their member states. Other institutions which are influential in free trade include:

- World Trade Organization (WTO): encourages free trade, polices free trade agreements, settles disputes and organises trade negotiations.
- International Monetary Fund (IMF): provides financial aid when a country has problems and promotes trade.
- World Bank: reduces poverty around the world by providing financial and technical assistance to developing countries.

Geopolitical relationships and changing patterns of development

Patterns of development have changed significantly since 1980.
Greater political cooperation means the world has become increasingly **interdependent** as a result of massively increased trade and cultural exchanges. The growth of newly industrialised countries (NICs) has been the most notable feature of this change. NICs are characterised by:

- rapid economic growth based on exporting goods or services
- rapid industrialisation and urbanisation
- stable governments and strong political leaders
- large amounts of foreign direct investment from overseas multinational corporations (MNCs)
- an expanding domestic market
- a well-developed infrastructure compared to other poorer countries
- the development of the NIC's own MNCs.

Example of a low-income country (LIC): Malawi

Figure 3 **A tealeaf picker in Malawi.**

In Malawi, agriculture employs 84 per cent of the labour force and accounts for 85 per cent of export earnings:

- The tea industry alone employs 50,000 seasonal workers. At the end of the season the tealeaf pickers are out of work.
- Tea is cultivated in 44 estates owned by eleven international companies.
- There are also 10,000 locally owned smallholder tea farmers in Malawi.
- Ninety per cent of Malawi's tea is exported to the UK and South Africa.
- Tea is processed in 21 factories; sixteen are owned by UK-based companies.
- The reliance on the export of agricultural commodities renders Malawi vulnerable to fluctuations in world trade, tariffs, quotas and environmental disasters such as drought.
- Malawi's imports are dominated by fuel products, comprising 30 per cent of its import bill.
- The economy is heavily dependent on aid from the IMF and the World Bank.

Example of a newly industrialised country (NIC): India

Reasons why India has experienced rapid economic growth include:

- Investment in education has provided a highly skilled workforce.
- Investment by MNCs such as IKEA and Samsung has created jobs in manufacturing.
- A stable government. India is the world's largest democracy.
- The second largest English-speaking workforce in the world.
- Well-developed infrastructure in terms of energy and transport.
- Low wage costs compared to competitors such as China and Mexico.
- Young workforce, flexible working practices and not heavily unionised.
- The world's second largest population. These people have purchasing power so the Indian market is large and attractive.
- Organisations like the World Trade Organization (WTO) promote free trade between countries, which helps to remove tariffs and quotas.

Figure 4 **Bagmane Technological Park, Bangalore.**

The transformation of the Indian economy has, unusually, been led by the service sector, particularly software and ICT services, together with media, advertising, retail, entertainment and tourism.

Revision activity

1. Sketch an outline map of India on a sheet of A5 card. Annotate your map with reasons for the growth of India as an NIC.
2. Locate and label Bangalore and Mumbai on your outline map.
3. Fill in your outline with ways in which you think the rapid growth of the economy will affect the lives of the Indian people. Add your card to your set of revision cards.
4. Carry out your own research and complete a revision card for either the Bagmane Technological Park or the Hindi movie industry.

Now test yourself

TESTED

1. Complete the following paragraph: Trade is the ... and import of goods and services between different countries. High-income countries usually sell manufactured goods and services whereas low-income countries sell ... goods such as ... An example of a low-income country is ... .
2. Name a newly industrialised country and give two reasons to explain why this country has developed rapidly in recent years.

The changing nature of global industry

The world is becoming increasingly interconnected. We communicate and share each other's cultures through travel and trade, transporting products around the world in hours or days in a huge global economy. **Globalisation** is the name of the process which leads to the free flow of goods, people, ideas and money on a worldwide scale, so that we increasingly live in a 'global village'. The drivers of globalisation include:

- improved technology, for example the internet allows e-commerce
- improvements in communications, for example the growth of satellite technology
- advances in transport, for example lower freight charges due to containerisation
- the removal of trade barriers, for example development of trading blocs
- the growth of MNCs.

The consequence of globalisation is a complex global web of interdependence.

Globalisation The process by which the world's economies, societies and cultures have become integrated through networks of communication, transportation and trade

Reasons why multinational companies locate plants in different countries

Companies that operate in several countries are known as multinational corporations (MNCs). The headquarters of an MNC is usually found in a 'global city' such as London. When money (capital) from one country is invested in another it is called foreign direct investment (FDI). Factors attracting investment in a country by a multinational include:

- government incentives
- availability of raw materials
- lower labour costs
- proximity to the markets where goods are sold
- ability to sell inside trade barriers
- reduced costs of buildings and land
- weaker legislation with regard to staff safety and welfare, protecting the environment and planning.

Revision activity

1. Explain why MNCs open factories in different countries around the world.
2. Complete a summary of Nike in Vietnam (see page 104) on a sheet of A5 card. Include:
   - reasons why Nike located in Vietnam
   - the good and bad things for the people who live in Vietnam (use different colours)
   - use the internet to print photographs of Nike products to add to your revision card.

What are the impacts of the location of MNCs in a country?

Advantages	Disadvantages
Investment provides new jobs and skills for local people	Profits are often sent back to the HIC where the MNC is based
Foreign direct investment brings foreign currency to local economies when they buy local resources, products and services	MNCs, with their massive economies of scale, may drive local companies out of business
The multiplier effect spreads the wealth through society	If an MNC finds it cheaper to operate in a different country, the factory may close and local people can be made redundant
	Lack of international laws means that MNCs may operate in a way that would not be allowed in an HIC, for example polluting the environment or imposing poor working conditions and low wages on local workers, perhaps using child labour

Example of Tata Steel in the UK

The Tata group is an Indian MNC. Its headquarters are in Mumbai. Tata owns 38 companies, including Jaguar Land Rover and Tetley Tea, and employs 50,000 people in the UK.

In 2006, Tata bought Corus, a large steelmaking company with plants in the UK. Buying European factories meant that Tata could sell steel more easily to European customers, avoiding tariffs. Competition from cheap Chinese steel has meant that Tata has considered closing its UK plants with the loss of 15,000 direct jobs and a further 25,000 indirect jobs in companies supplying Tata with parts and services. The result could be to send parts of south Wales and Yorkshire into a spiral of economic decline – a negative multiplier.

Example of Nike: a multinational corporation

Nike was founded in Oregon, USA, in January 1964:

- It has factories manufacturing sports equipment in 41 different countries and sells goods in over 700 stores worldwide.
- Production is **outsourced** to countries such as Indonesia, China, Taiwan, India, Thailand and Vietnam.
- Global sales in 2014 were $28 billion.

Figure 5 A sports shoe factory in Vietnam.

Nike in Vietnam, a newly industrialised country (NIC)

Vietnam is a **newly industrialised country (NIC)** in South-East Asia. Over 75 million pairs of shoes are made annually for Nike in Vietnam. Nike located in Vietnam because of:

- low labour costs
- less expensive factories and land
- low tax rates
- low energy bills
- fewer laws and restrictions, for example on workers' rights.

Outsourcing Transferring of work to outside suppliers rather than completing it internally

Newly industrialised country (NIC) A country in which development has been rapid over its recent history, often greater than seven per cent a year

Impacts of Nike in Vietnam:

Advantages	Disadvantages
Created 40,000 jobs and improved the skill base of the local population	The company image and its advertising may help to undermine national culture
Pays higher wages than most local companies	There are concerns about the political influence of Nike in Vietnam
Helped to attract more MNCs	Investment could be transferred quickly from Vietnam elsewhere, leaving people without jobs
It contributes to tax, which helps to pay for education and infrastructure	Factories gained a reputation for sweatshop conditions. There have been stories of abuse of workers
Export of Nike products brings money into the country	Nike has a large demand for energy and water

Exam practice

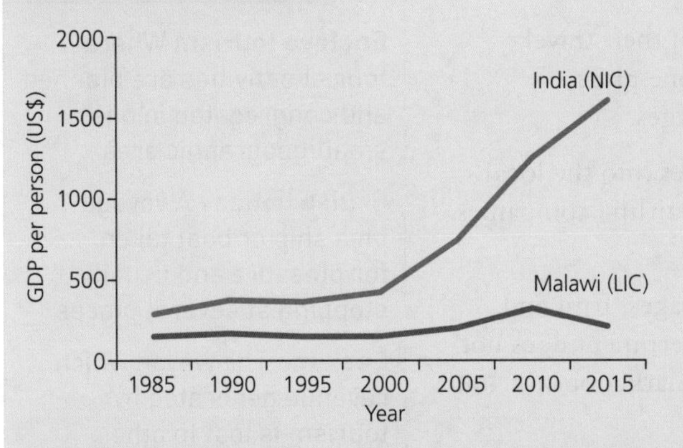

Figure 6 GDP per person for Malawi and India.

1. Suggest one alternative way of representing the information shown in the graph above. Justify your choice. [2]
2. Compare the change in the GDP per person for Malawi with that of India [2]
3. Suggest ways in which this graph could be misleading. [4]

Exam tip

Twenty-five per cent of the marks in the examination will test skills. Research the WJEC website, study pages 33 and 34 of the Geography GCSE specification or pages 30 and 31 of the Eduqas Geography A GCSE specification. Make a list of the skills you need to know about. Carry out a traffic lights exercise to check your knowledge and understanding.

The growing contribution of tourism to the globalised economy

Tourism accounts for one in eleven jobs worldwide. Until recently it was an activity which was concentrated in the richer countries of the world but it is increasingly becoming an important source of income in low- and middle-income countries.

Key drivers of the globalised tourist industry

	Tourist arrivals by year (in millions)		
Region	**1995**	**2014**	**2030**
Africa	18.7	55.7	134
Middle East	12.7	51.0	149
Americas and the Caribbean	109.1	181.0	249
Asia and the Pacific	82.1	263.3	535
Europe	304.7	581.8	744
Total	527.3	1132.8	1810

Figure 7 The growth of the global tourist industry – the number of tourists in 1995 and 2014 and the predicted number in 2030.

The age of mass tourism began in the 1960s. Factors that have contributed to the rapid growth in tourism include:

- Increased life expectancy and early retirement mean that there are a greater number of older people travelling.
- Higher salaries and savings mean that people have more spending money.
- Modern aircraft mean that travel has become easier and airfares have become more affordable.
- The growth of holiday companies has made booking easier and holidays more affordable. Tourists often purchase a package that includes flights, hotels and meals.
- The internet has allowed people to find out about holiday destinations and make their own travel and accommodation arrangements.
- More stable governments around the world have led to easier and safer travel.

Now test yourself

1. Describe one named change in technology that has allowed increased globalisation of human activities.
2. 'Investment by multinational corporations in a developing country is always a good thing.' How far do you agree with this statement in relation to one MNC that you have studied?
3. Explain how investment by companies such as Nike into developing countries is changing global patterns of development.

TESTED ☐

Tourism Any activity where a person voluntarily visits a place away from home and stays there for at least one night

Enclave tourism and the informal economy

Features of recent developments include:

- **Enclave tourism**: tourists pay one price and get all of their travel, accommodation, food, drink and entertainment in one place.
- **Cruise holidays**: cruise ships sell 'all-inclusive' packages.

In such holidays, less of the money spent by tourists goes into the local economy, it is **leaked** back to the MNCs that own the airline companies, hotel chains and cruise ships.

Many tourist-related activities, for example beach massages, fruit and souvenir sellers, work in the informal sector so the government does not earn any direct income. This is described as the black market or **informal economy**.

As wages rise in NICs, so the demand for tourism is growing in the indigenous populations. For example, about 100 million tourists travelled abroad from China in 2014, further fuelling the global growth in tourism.

Enclave tourism Where tourist activities are planned and congregated in one small geographic area

Cruise holiday A voyage on a ship or boat taken for pleasure and usually stopping at several places

Leakage The way in which revenue generated by tourism is lost to other countries' economies

Informal economy The jobs that are done by self-employed people which are neither declared to, nor regulated by, the authorities

Impact of tourism's growth on employment structures

Many countries see the promotion of tourism as a development strategy. Tourism not only creates growth in the tertiary sector, but can also have a positive multiplier effect and causes growth in the primary and secondary sectors as well. Money spent in a hotel helps to create jobs directly in the hotel, but it also creates jobs indirectly elsewhere in the economy. The hotel, for example, buys food from local farmers, who may spend some of this money on fertiliser or clothing. The demand for local products increases as tourists often buy souvenirs, which increases secondary employment.

The impacts of the growth of the tourist industry include:

Benefits	Costs
Generates employment	Jobs are often low paid and temporary
Brings foreign exchange	Local culture is destroyed
Wealth can be invested in services such as health and education	Fragile ecosystems, for example sand dunes, are destroyed
New facilities provided for tourists can be used by locals	Increased pollution from road and air traffic

Now test yourself

TESTED

1. Describe how improvements in technology and transport have helped the growth of a globalised tourist industry.
2. Explain why local economies have limited benefits from 'enclave tourism'.

Revision activity

1. Complete a spider diagram summarising the reasons for the growth in tourism since the 1960s.
2. Use Figure 7 to describe and compare the growth of tourism in different regions of the world. Can you suggest a reason why tourism is growing more rapidly in some regions?

Example of tourism in an NIC: Vietnam

Over 6 million tourists visited Vietnam in 2015. Reasons for the increase in numbers of tourists include:

- Improved transportation: especially air travel.
- The Communist government has deregulated the economy to allow greater private ownership and relaxed visa rules to make it easier to visit the country.
- Better image: the Vietnam War finished over 40 years ago.
- Attractive foreign exchange rates make Vietnam a relatively cheap place to visit.
- Unique human and physical tourist attractions.

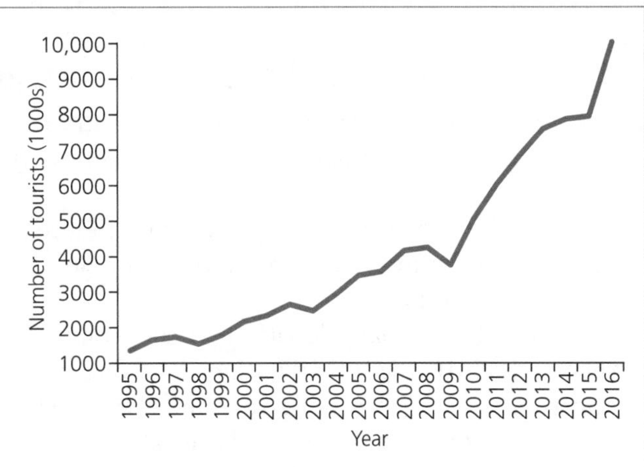

Figure 8 Tourist arrivals in Vietnam, 1995–2016.

Vietnam's tourist attractions include:

- Cu Chi Tunnels: used during the Vietnam War by the Vietcong.
- Temples: a large number of religions are practised in Vietnam including Buddhism and the unique Caodaism.
- Cuisine: Vietnam has a wide variety of different foods.
- Halong Bay: spectacular scenery with thousands of limestone karsts.
- Lang Co Beach: a beautiful spit that has a peaceful lagoon behind it.
- Wildlife: diverse and interesting.

Impacts of tourism in Vietnam include:

Benefits	Costs
Employs about 250,000 directly and 500,000 indirectly, bringing $16.4 billion to the economy in 2015	Many tourist developments are partly owned by foreign companies. Some profits leak overseas
Tourism is encouraging greater entrepreneurship and improving language skills, for example tour guides and taxi drivers	Jobs are seasonal, many people lose their jobs during the wet season when fewer visitors arrive
New services are built and developed which benefit tourists and local residents, for example hospitals	A serious issue is the growth of sex tourism, especially the recruitment of young boys and girls
New national parks are being created to protect wildlife	The arrival of tourists can cause a decline in local cultures, for example loss of language

Now test yourself

TESTED

'The growth of tourism is good for the development of lower-income countries.' To what extent do you agree with this statement?

Exam practice

1 Draw a suitable graph to show the information in the table below. Justify your choice of graph. [4]

Cruise passengers worldwide (millions)	4.0	5.5	8.0	11.5	16.5	25.0
Year	1990	1995	2000	2005	2010	2015

2 Describe the trend shown by your graph. [2]
3 Give two reasons to explain this trend. [4]

Revision activity

1 Describe the growth in tourism in Vietnam.
2 Research the internet and find and print photographs of the tourist attractions. Now print an outline map of Vietnam. Locate the attraction on your map and add your photograph of the attraction.
3 Annotate your map with reasons for the growth of tourism in Vietnam.

What are the responses to uneven development at the global scale?

REVISED

How international aid can reduce inequality

Aid is a transfer of resources from a richer country to a poorer country. It includes money, equipment, training and loans. Different types of aid include:

- Bilateral aid: between two countries, it is often tied aid, meaning that the receiving country must spend money on goods and services from the donor country.
- Multilateral aid: money donated by richer countries via organisations such as the IMF, the **United Nations (UN)** and the World Bank.
- Short-term emergency aid: gives immediate relief during or after a natural disaster such as drought; it includes food, medicines and tents.
- Long-term development aid: a sustained programme of aid which aims to improve standards of living, for example education for young people.
- Debt abolition: when richer countries cancel debt owed to them by poorer countries.
- Aid from non-governmental organisations (NGOs): given through charities such as Oxfam and Save the Children.

United Nations (UN)
An intergovernmental organisation to promote international cooperation

Intermediate technology
Low-cost, often labour-intensive technology, based on local resources, that is appropriate to economically less developed countries

Example of aid helping to reduce inequality in an LIC: Malawi

Most people in Malawi are subsistence farmers. The area in Middle Shire is affected by soil erosion during the rainy season. Reasons for this include:

- A rapidly rising population means the farmers have cut down forests to provide land for growing food and for firewood.
- Tobacco is the main cash crop; falling prices mean that more needs to be grown to get the same return, resulting in more intensive cultivation.

The Community Vitalization and Afforestation in Middle Shire (COVAMS) is a ten-year project funded by the Japanese government. The project aims to prevent soil erosion by:

- education on the causes of soil erosion
- training farmers to plough around the hillsides, following the contours and slowing surface runoff
- building rock, wood and bamboo barriers across streams to slow down the flow of water
- building terraces into the hillsides to reduce surface runoff
- supplying fast-growing tree species from local nurseries to speed up re-afforestation.

This approach uses **intermediate technology**, local skills and local materials to provide low-cost solutions to the problem.

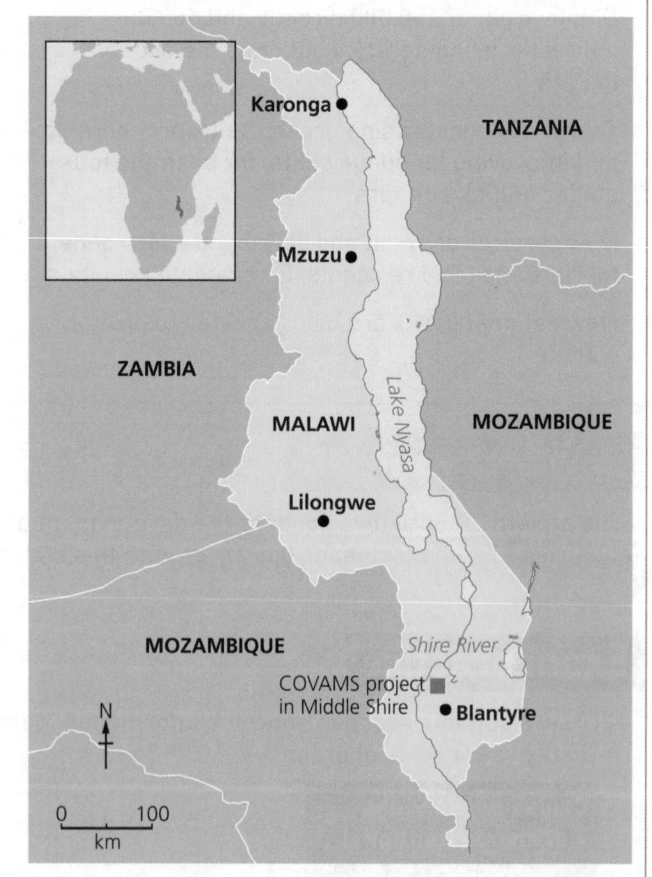

Figure 9 Location of Malawi.

Large-scale projects 'kick start' the development process. This is a top-down approach often funded by organisations such as the World Bank, for example building a dam to provide hydroelectric power. Small-scale projects work with local people and use local skills. This can be described as bottom-up development often supported by NGOs, for example digging wells to provide clean water.

Revision activity

On a sheet of A5 card draw a sketch map to show the location of the COVAMS project in Middle Shire. Annotate your map to show how aid is helping the farmers of Malawi.

Development targets

The UN is supported by 192 countries. One of its aims is to encourage human development. To help achieve this goal, in 2015 the UN set seventeen Sustainable Development Goals to be achieved by 2030. These build on the Millennium Development Goals (MDGs), which ran from 2000 to 2015. To find out more about the work of the UN visit its website: **https://sustainabledevelopment.un.org/?menu=1300**

The benefits of fair trade

Fair trade means that the producer receives a guaranteed and fair price for a product. It aims to provide:

- minimum wages and safe working conditions
- restrictions on child labour
- protection of the environment
- improved schools and healthcare.

Fair trade products sometimes cost a little more, but many consumers consider this a small price to pay for the benefits they bring. Fair trade products include tea, coffee, sugar, chocolate and clothing made from fair trade cotton.

Commodities Raw materials or primary agricultural products that can be bought and sold, such as tea or copper

Example of fair trade: Ghana's cocoa trade

Ghana is a country in west Africa. Trading in gold, oil and cocoa has helped Ghana to develop in recent years:

- 2.5 million farmers grow cocoa as their main crop
- 90 per cent is grown on smallholdings, tiny farms smaller than 3 ha in size
- most cocoa is sold for export; only about five per cent of the cocoa crop is processed into chocolate in Ghana
- 75 per cent is exported to the EU where it is made into chocolate in countries such as Belgium, Germany and the UK
- production fluctuates from year to year depending on the weather, pests and diseases
- the average income for a cocoa farmer is only about £160 a year.

Cocoa and crops such as tea and sugar are examples of **commodities**. Traders in Europe buy cocoa beans in a futures market in London. The price can go up or down from day to day, depending on supply and demand. The fluctuating price makes it very difficult for farmers in Ghana to earn a fair wage.

Fair trade is helping small-scale farmers in Ghana:

- Day Chocolate Company was set up in 1997 by Kuapa Kokoo, a farmers' cooperative, with the support of the Body Shop and Comic Relief, to produce a local chocolate bar.
- Divine Fairtrade milk chocolate was launched in October 1998. The cocoa farmers receive a share of the profits from the sale of Divine as well as a fair price for their cocoa beans.

Now test yourself

TESTED

1. What are commodities?
2. Outline two advantages of fair trade for cocoa growers.

Arguments for and against aid

For	Against
Emergency aid saves lives and reduces misery	Aid can increase dependency on the donor country
Development projects lead to long-term improvements in standards of living	Profits from large projects can go to multinationals and donor countries
Assistance in developing natural resources benefits the global economy	Aid doesn't always reach the people who need it and can be kept by corrupt officials
Aid for industrial development creates jobs, and aid for agriculture helps to increase food supply	Aid can be spent on prestige projects or in urban areas rather than in the areas of real need
Provision of medical training and supplies improves health	Aid can be used as a weapon to exert political pressure on the receiving country

Revision activity

1 Put the reasons in favour and the reasons against giving aid on relevant sides of a set of scales. What do you think? Is aid making a difference to the lives of people in poorer countries?
2 List some advantages for the donor country of giving aid.

Exam practice

Discuss how receiving international aid could improve the lives of people who live in the least developed countries. [8]
The accuracy of your writing will be assessed in your answer to this question. [3]

Exam tip

Writing accurately will be assessed in certain questions that require extended writing. Writing accurately takes into account the use of specialist geographical language and the accuracy of spelling, punctuation and grammar.

Questions such as the one above are worth a total of 11 marks out of the 80 for the entire paper. It is important therefore that these questions are given particular attention.

Water resources and their management

How and why is the demand for water changing?

REVISED

Global trends in water consumption

The global consumption of water has been rising over time. The two main reasons for this are:

- Increasing population: people use water for drinking, hygiene, cooking and cleaning. Water is also needed to produce the food we eat and make the goods we use.
- Economic development: the **water footprint** of HICs is much higher than LICs. As more countries develop the global demands for water will increase.

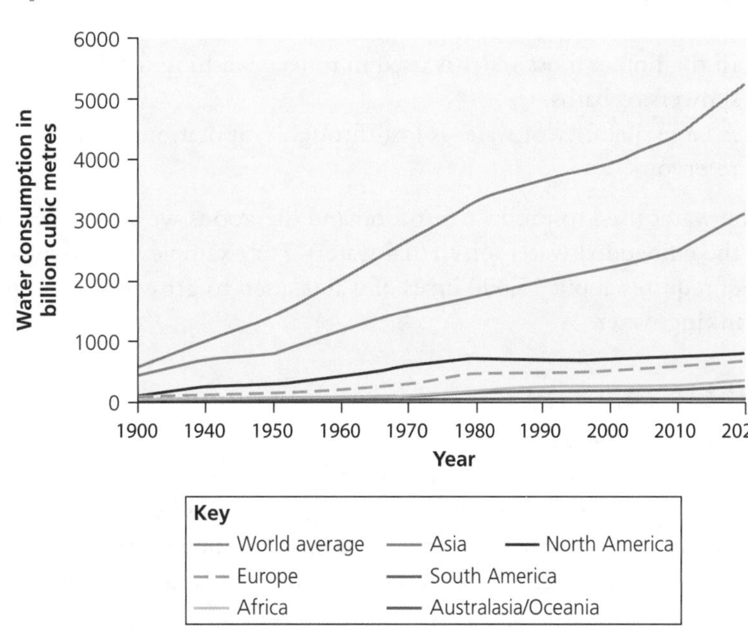

Figure 10 Water consumption worldwide, 1900–2025.

The consumption of water varies within, as well as between, countries. In sub-Saharan Africa people living in urban areas are twice as likely to have safe, piped water as those living in rural areas. In rural areas, girls spend hours walking to collect water rather than attending school. In urban areas people living in shanty towns often have to buy water from street vendors and can pay up to 50 times the amount for water paid by people living in European cities.

Water footprint The total volume of freshwater consumed and polluted by people. It is calculated by adding the direct and indirect water use of people

Revision activity

1. Use Figure 10 to describe how global water consumption has changed since 1900.
2. Suggest reasons to explain why the average American uses 1300 litres of water a day compared to the average African using only 22 litres a day.
3. Use evidence from the photograph below to suggest how building a well would encourage 'bottom-up' development in the Sahel.

Figure 11 African women of the Sahel spend a lot of time doing work that does not contribute directly to family or state income.

Now test yourself

Suggest reasons which would explain why water consumption has increased more in Asia than in any other global region.

TESTED

Water footprints

A person's water footprint measures the total amount of water used in everyday life for drinking, cooking and washing, together with the water that is used to grow food and to produce goods and services. The water footprint tells us how much water is being consumed by an individual, a particular country or globally:

- 70 per cent of the water we use is to produce our food, 20 per cent is used by industry and 10 per cent is used in the household.
- In agriculture, the majority of water is used for irrigation.
- Industry uses large quantities in manufacturing and cooling processes.
- In the home, most water is used in toilets, washing machines and showers or baths.
- A large quantity of water is lost through evaporation and leaks in reservoirs.

The water used to produce our food and the goods we purchase is known as the embedded water (or virtual water). For example, to produce 1 kg of beef requires about 15,000 litres of water, used to grow grass and provide drinking water.

Water security

Water security is when:

- people have enough safe and affordable water to stay healthy
- there is sufficient water for agriculture and industry
- the supply is sustainable and ecosystems that supply water are conserved
- people are protected from water-related hazards such as drought.

Economic water scarcity is when water is available, but for some reason it is inaccessible or unusable. This might be because it is groundwater that is expensive to extract or that the cost of transporting it is too expensive or that the supply of water has become polluted.

Physical water scarcity is when there is not enough water available. The most common reason for this is low precipitation rates.

Reasons for increasing water demands

- Population growth: in 1900 the global population was 1.6 billion, today it is over 7 billion and it is forecast to be over 10 billion by the end of the century.
- Agricultural change: in richer countries irrigation is often mechanised, with sprinklers using vast amounts of water.
- Industrial growth: in richer countries this is often on a large scale with industries such as steel demanding huge quantities of water. Poorer countries generally have smaller-scale cottage industries which demand less water. As more multinational companies invest in NICs and LICs, there will be a greater demand for water, for example in India drinks manufacturers use over a million litres of water a day.
- Growth of **consumerism**: in richer countries there are many facilities which demand water, for example showers, dishwashers and washing machines. Water is also used in the leisure industry, for example swimming pools, spas and golf courses. In poorer countries, many people do not have access to piped water and so use it more sparingly.

Water security When the population of a country has sustainable access to adequate quantities of acceptably clean water

Consumerism An ideology that encourages the acquisition of goods and services in ever-increasing amounts

Revision activity

Try to estimate your daily water footprint. You may find websites that will help you do this.

Now test yourself

Define 'water security', water footprint and consumerism.

TESTED

Exam practice

Give two reasons to explain why the consumption of water has increased globally. [4]

Are water resources being managed sustainably?

REVISED

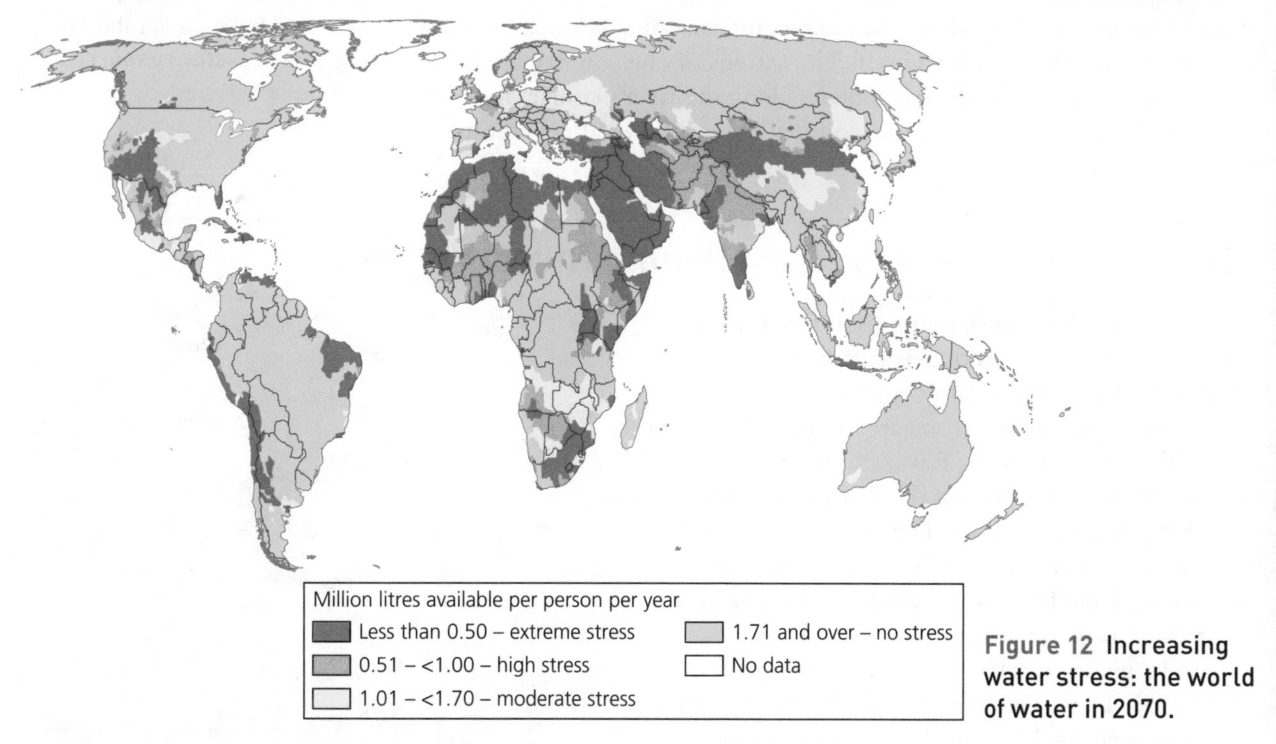

Figure 12 Increasing water stress: the world of water in 2070.

The map above shows a projection of per capita water availability in the year 2070. Some parts of the world are more at risk of water shortages than others. As populations increase and countries develop the number of places suffering from **water stress** is likely to increase. For more than 748 million people around the world, safe, clean water is a luxury. Climate change is likely to have an impact on the number and distribution of places suffering from water stress in the future.

Water stress When the demand for water exceeds the available supply or when poor quality of water restricts its use

Strategies to manage water supply

- Building dams and reservoirs: this provides large supplies of water all year round for domestic use and irrigation.
- Water transfer schemes: when a country has a water surplus in one area and a water shortage in another, supplies can be transferred through canals and pipes.
- Desalination plants: removing the salt from seawater to make it safe to drink. The process is expensive and uses large amounts of energy.
- Abstraction of groundwater: water found underground in the cracks and spaces in soil, sand and rock. It is stored in rock formations called aquifers.
- Water conservation: examples include dual-flush toilets and water meters. At a national level, water companies try to conserve water by fixing leaking pipes.
- Using 'grey' water: water that has either been used previously or is untreated rainwater. Some uses of water do not require water that has been purified, for example flushing toilets.

Revision activity

Use Figure 12 to describe areas of the world that are predicted to be under extreme water stress in 2070.

Now test yourself

1. Describe two ways in which the supply of water can be managed to meet increasing demands.
2. What is rainwater harvesting and how could this reduce domestic demand for water?

TESTED

Consequences of water management at an international scale

Water politics (**hydropolitics**) is politics that is affected by the availability of water resources. Many rivers flow across international boundaries and sometimes form the boundary itself. The actions of one country are likely to have consequences for other countries along the river's course, for example if water is polluted in one country then this will affect all other countries downstream.

Hydropolitics Politics affected by the availability of water and water resources

Example of water politics: Mekong River, Vietnam

Two-thirds of Vietnam's water resources originate outside the country, making Vietnam susceptible to decisions made in upstream countries:

- Vietnam's two major rivers are the Red River in the north and the Mekong River in the south.
- The Mekong flows from the Tibetan Plateau through China, Myanmar, Laos, Thailand and Cambodia, before reaching the sea via the Mekong Delta.
- The Mekong Delta has been described as a biological treasure trove and the rice bowl of Vietnam, producing over half of the country's staple food.
- In 1995, Laos, Thailand, Cambodia and Vietnam established the Mekong River Commission.
- In 1996, China and Myanmar joined discussions to manage the Mekong's resources.

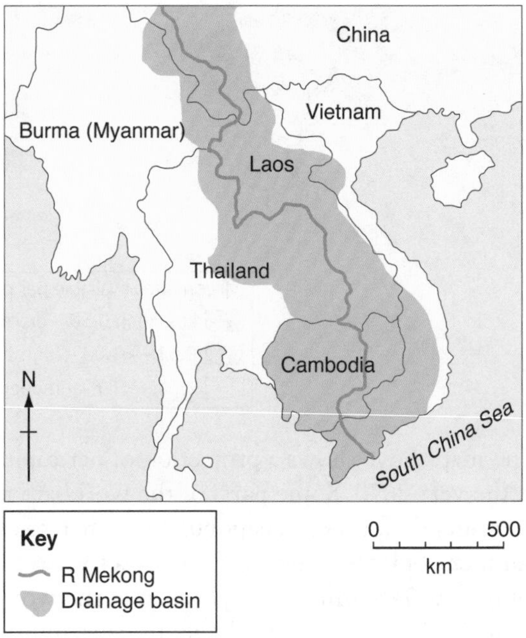

Figure 13 Location of the Mekong River.

The aims of the Mekong River Commission include:

- Achieving closer co-operation with China, discussing issues such as the effect of building dams in China, maintaining a dry season flow and giving early flood warnings.
- Expanding areas of irrigated farming.
- Developing drought strategies to ensure greater reliability of water supply.
- Protecting ecosystems.

Consequences of small-scale water management

Small-scale water management projects are often funded by **NGOs**. They usually:

- are bottom-up projects controlled by the local community
- are relatively cheap and easy to set up
- are easy to maintain using simple or intermediate technology
- address local issues.

NGO A non-governmental organisation (NGO) is a not-for-profit organisation that is independent from states and international governmental organisations

Example of a small-scale project: WaterAid in Malawi

In Malawi, the number of people with reliable access to clean water is much lower than the official figure of 90 per cent. Many hand pumps are broken and in rural areas as few as two in ten people have access to a toilet.

WaterAid supports communities to repair broken hand pumps, improve hygiene practices and train people to maintain their own resources.

In villages without a toilet, WaterAid helps to build simple composting toilets that keep water sources clean and also provide fertiliser for crops.

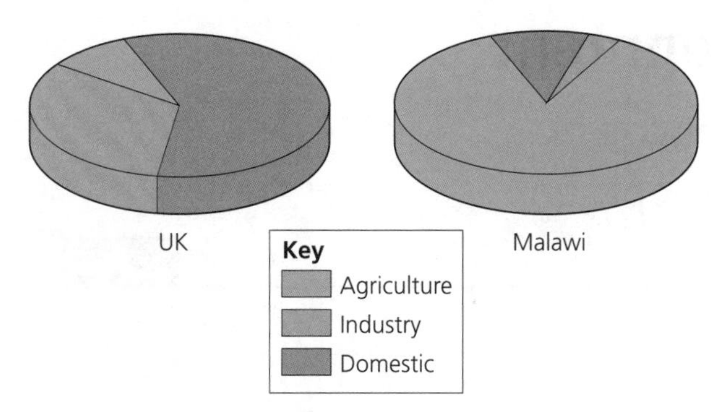

Figure 14 Water use in Malawi and the UK.

Revision activity

1. Make a list of countries through which the River Mekong flows.
2. Why is international management of the Mekong River and its tributaries important?

Now test yourself

TESTED

1. Compare water use in Malawi to the UK.
2. Explain why Malawi uses a much greater percentage of its water on agriculture.
3. Suggest reasons to explain why it is often difficult to achieve international agreement.

Example of over-abstraction of groundwater: India

In India, groundwater provides 65 per cent of all water used in agriculture and 85 per cent of domestic water use. This water is stored in porous rock known as an **aquifer**. In some areas, over-abstraction is taking place where water is taken from the ground faster than it can be recharged through rainfall percolating down into the ground. In many places the **water table** is 4 m lower than it was in 2000.

Reasons for the use of groundwater include:

- Some states, such as Gujarat, have a long dry season. Rivers and lakes dry up and groundwater is the only source of water.
- Surface stores are often polluted; groundwater is considered cleaner.
- Cheap electricity has encouraged farmers to drill deep wells and install pumps to extract water.
- The Green Revolution has led to farmers growing crops that need more water than traditional crops.

In the future, the Indian government is going to have to find alternative ways of supplying the growing demand for water:

- The government could build more dams. India already has 3200 major- and medium-sized dams, and building more would be very unpopular since more villages and farmland would be flooded. This is an example of a top-down approach to the problem.
- A bottom-up approach would be to encourage farmers to adopt water conservation measures. One example is **rainwater harvesting** where farmers build earth dams to create small pools of water which fill during the rainy season. In the following months this water can be used and will slowly sink into the ground, recharging the aquifer.

Aquifer A permeable rock which stores and transfers water

Water table The upper boundary of the underground saturated part of the soil or rock

Rainwater harvesting The collection of water from surfaces on which rain falls, and storing this water for later use

Exam practice

Explain why the demand for water is higher in high-income countries than in low-income countries. [6]

Exam tip

Always look to give specific examples in your answer. In response to this question the examiner would expect you to refer to specific HICs and LICs to gain full marks.

Regional economic development

What are the causes and consequences of regional patterns of economic development in India?

REVISED

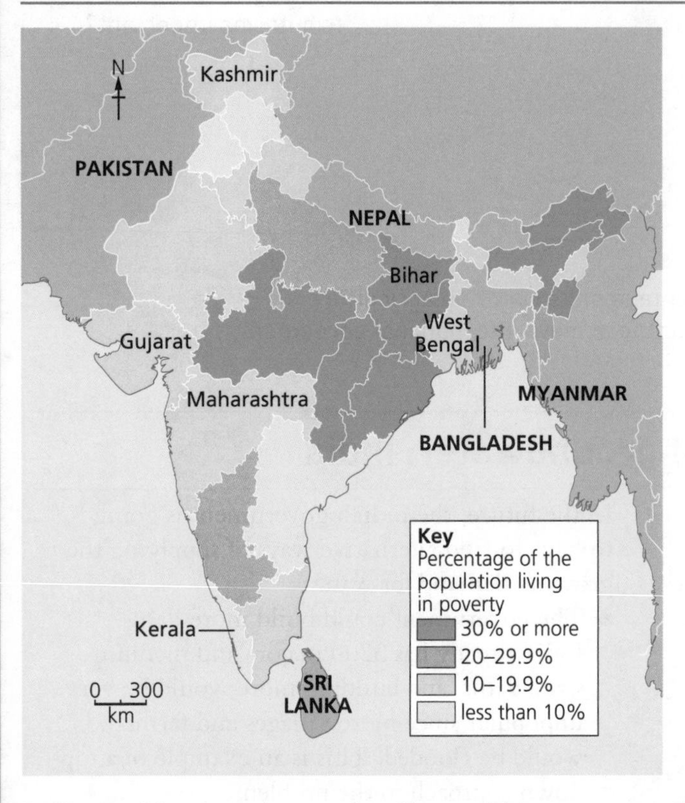

Figure 15 **Percentage of the population of India living in poverty, 2013.**

India provides a good example of how within a country there are regional differences in development. There are huge inequalities between the richest and poorest members of Indian society. Patterns include:

- In Kerala, the state government has always funded public education and health. The result is low birth rates, high life expectancy and relatively high quality of life.
- Kashmir has lower than average levels of per capita income. This can be attributed to conflict in the region between India and Pakistan.
- Maharashtra and Gujarat are India's main centres of heavy industry, producing 39 per cent of the country's industrial output and 67 per cent of its petrochemicals.
- The north-central region is poor although West Bengal stands out for its relatively high income levels. It is noted for its heavy industries and it has also long been one of India's main intellectual centres.

India also has climatic differences, ranging from arid desert in the north-west, tundra in the Himalayas to the north, and humid tropical regions supporting rainforests in the south-west. These differences have major impacts on farming in particular.

While it has been illegal in India to discriminate against others based on caste since the 1950s, the caste system continues to affect society in terms of economic inequality and even the election of politicians.

Discrimination against girls and women is widespread in India. The dowry system (payment of money or goods by the family of a bride) means that girls are a financial burden, which further promotes gender inequality.

Example of regional social and economic inequalities: Bihar vs Maharashtra

In 2010, the average per capita income in Bihar was £251 per year but in Maharashtra it was £1011. Maharashtra has initial advantages which include fertile soil, a good climate for farming, a good water supply and a coastline which makes trade and transport easy.

Bihar	Maharashtra
80 per cent of the population live in rural areas; it is part of India's **rural periphery**	Has three of India's largest cities, Mumbai, Pune and Nagpur; it is part of India's **urban core**
Education is poor and the birth rate is high	It is a centre for banking, insurance and call centres
Many people work as landless farm labourers	It has manufacturing industries, for example steel and textiles
Farms are small and produce little more than a family needs to survive	Mumbai is the centre for India's film industry in Bollywood
Government is more corrupt than elsewhere in India	Government jobs in the region are well paid

The consequence of inequality is protests, the growth of extremist movements, demands for regional independence and the exclusion of 'non-natives'. Crime is increasing in some areas and there is widespread public anger at political corruption as well as ways in which state policy has favoured the rich. These forces create sources of instability that may harm the development process in India in the future.

Rural periphery On the edge or margins; areas which have a poor economic status and thus suffer from the associated social conditions

Urban core An area that enjoys economic, social and political superiority in comparison to its surrounding area

Now test yourself

TESTED

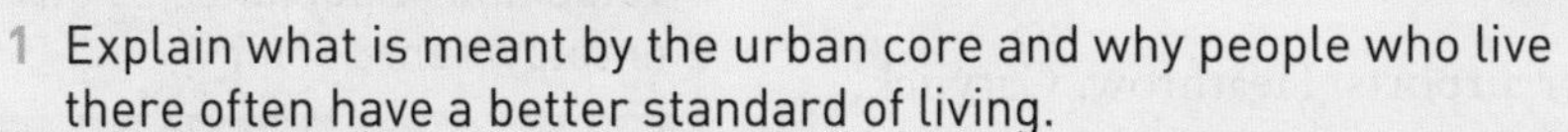

1. Explain what is meant by the urban core and why people who live there often have a better standard of living.
2. Complete a flow diagram to show a predicted multiplier effect from a government investing in education and health, as in Kerala.

Revision activity

1. Use Figure 15 to describe the distribution of areas where more than 30 per cent of the population live in poverty.
2. Identify one social, one economic, one cultural, one political and one environmental reason for regional differences within India.

Exam practice

State	Percentage of population living in poverty
Bihar	34
Kerela	7
Maharashtra	17

1. Display this information using a suitable graph. [4]
2. Suggest two reasons to explain why a large percentage of the population of Bihar in India live in poverty. [4]

What are the causes and consequences of regional patterns of economic development in the UK?

REVISED

The north–south divide

In the UK, it has long been recognised that there is a north–south divide. This refers to the social and economic differences between southern England and the rest of the UK. Reasons for these differences include:

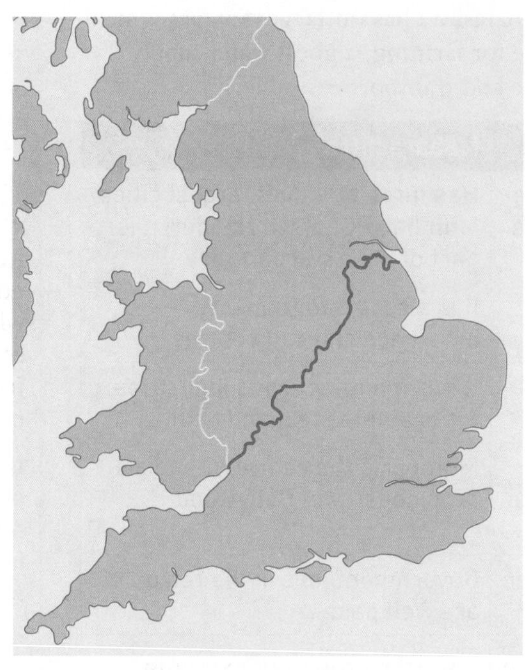

Figure 16 The UK's north–south divide.

- The zone stretching along the M4/M11 motorway corridor between Bristol, London and Cambridge has attracted modern high-tech manufacturing industries.
- Research and development spending in the south-east in 2010 was £3.6 billion compared to £0.3 billion in the north-east.
- The south has a more educated workforce with higher paid jobs in science, technology and finance.
- London is a global centre for banking and finance.
- The south-east has many motorway networks concentrated around it, making it easier for firms to distribute their products around the country or world.
- The south is home to the UK's major airports: Heathrow, Gatwick, Stansted, Luton and Bristol.
- The north was once heavily industrialised but these manufacturing industries have declined due to **deindustrialisation**, for example the steel industry has lost over 200,000 jobs since 1967.
- In political terms, the south is largely supportive of the Conservative Party, while the north is generally more supportive of the Labour Party.

Deindustrialisation Fall in the percentage contribution of secondary industry to an economy in terms of its value and importance as an employment sector

Now test yourself

Give three reasons to explain why there is a north–south divide in the UK.

TESTED

Social and economic consequences of regional inequalities

Consequences include:

- Health conditions are generally poorer in the north.
- Life expectancy is higher in the south.
- House prices are higher in the south, particularly the south-east.
- Incomes are higher in the south.
- There is a migration of young professionals from the north to work in the south and London in particular.

Exam practice

Name the process which results in a growing range of economic activities. Underline the correct answer from the following words: deindustrialisation, regeneration, diversification. [1]

Example of Sheffield: a city in the north

Sheffield was once an important centre in the UK's steel industry. In the 1980s the industry declined dramatically: 120,000 people lost their jobs between 1971 and 2008.

The deindustrialisation Sheffield experienced led to economic **diversification** and the growth of jobs in retail, software development and business services. Deindustrialisation has brought many environmental and social benefits but at great economic cost:

Positive impacts	Negative impacts
Water quality: during the industrial period rivers and streams were polluted with industrial waste. Now rivers, such as the Don, have much improved water quality and are being restocked with fish	**Derelict land:** about 900 ha of derelict land and buildings were left. Much of the land is polluted with heavy metals and other industrial waste
Air quality: air quality in the city has improved	**Greenfield sites:** as people are unable to find work in inner Sheffield there is pressure on the greenfield sites on the edge of Sheffield for homes
Regeneration: some of the old industrial sites are now available for regeneration, for example Hadfield's steelworks is now Meadowhall shopping centre	**Traffic:** as jobs in Sheffield are becoming scarcer, people are having commute to reach their jobs, increasing congestion

How can regional inequalities in the UK be reduced?

REVISED

How investment creates growth in deprived areas

Successive governments have developed regional policies to try to reduce inequalities within the UK:

- The Regional Growth Fund (RGF) was created in June 2010 with the intention of promoting the private sector in areas in England most at risk of public sector cuts. The RGF was worth over £3.2 billion and ran until March 2017.
- Enterprise Zones are designated areas across England that provide tax breaks and government support. First established in 2012, 48 Enterprise Zones were in place by April 2017.
- Enterprise Zones in Wales offer incentives to attract new business to prime locations. Their objective is to grow the local economy, provide new jobs and act as a catalyst for growth.

Diversification The process which results in a growing range of economic activities

Example: Sheffield City Region Enterprise Zone

This is located across six sites along the M1 motorway corridor, in the centre of the UK, with London 90 minutes away by train. The zone is home to the University of Sheffield Advanced Manufacturing Research Centre (AMRC), a world-class centre for research into advanced manufacturing technologies used in the aerospace, automotive, medical and other high-value manufacturing sectors.

Positive and negative multipliers

One way to tackle regional inequality is through investment in major **infrastructure** projects, such as road and rail schemes. Better transport links will attract new industries and lead to a positive economic multiplier.

Infrastructure
The collective name for all the communication links and basic utility links that are built across a country to facilitate movement

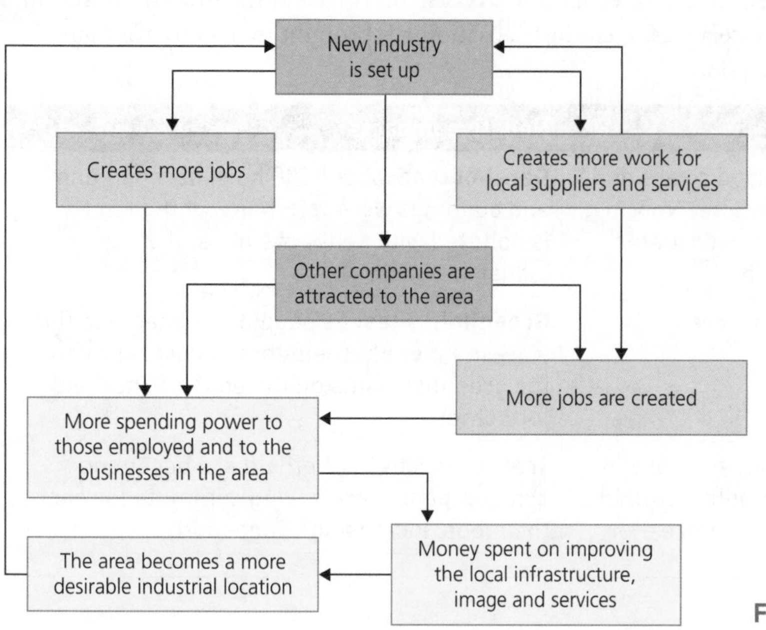

Figure 17 Positive economic multiplier.

Example: relief road around Newport

The M4 motorway is the main transport route that runs through south-east Wales and links the region with London and south-east England. Between Newport and Cardiff, the motorway narrows to two lanes as it crosses the River Usk and through the Brynglas Tunnels. This causes congestion and major traffic hold-ups. In 2014, the Welsh government confirmed its intention to build a £1 billion relief road. This is the largest capital investment programme ever announced by the Welsh government.

Many cities in the north, such as Sheffield, have experienced a negative multiplier. The closure of industry means that people are unemployed and do not have money to spend in local shops.

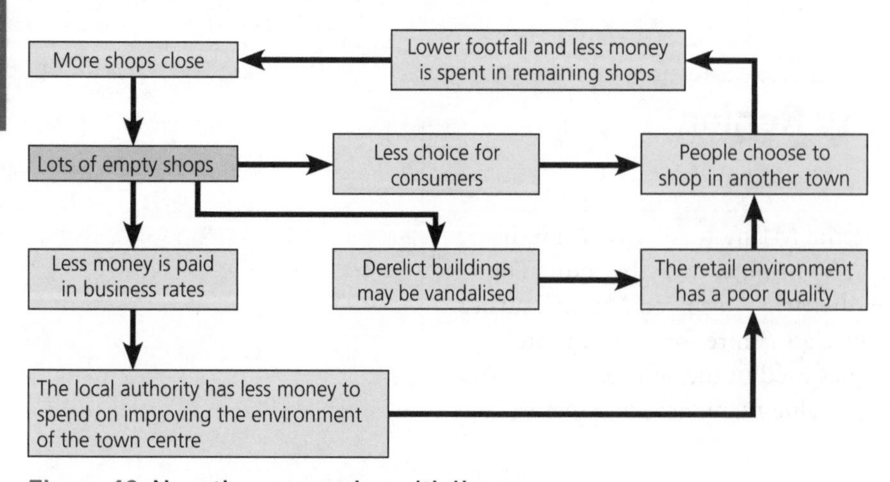

Figure 18 Negative economic multiplier.

How national policies reduce regional inequality

National policies include:

- The government is devolving power to local councils in the north, for example the election of the mayor for Greater Manchester in May 2017.
- The creation of the 'Northern Powerhouse' which is a proposal to boost economic growth, particularly in the core cities of Manchester, Liverpool, Leeds, Sheffield and Newcastle. This would be an attempt to turn the north's population of 15 million into a collective force and to attract investment into northern cities and towns.
- The improvement of transport links such as the High Speed 2 train which will connect London with Birmingham, and in a later phase, cities in the north including Manchester, Sheffield and Leeds. Better transport links will attract new investment and create an economic multiplier.
- The relocation of some businesses and organisations, for example the BBC built MediaCityUK near Manchester and moved many of its offices there in 2011. Since then, the multiplier effect has led to other companies locating close by, for example the Holiday Inn hotel chain.

Now test yourself

TESTED

Explain how deindustrialisation in areas such as Sheffield and the south Wales valleys can lead to a negative multiplier.

Revision activity

Create a flow diagram to show a possible positive economic multiplier following the construction of the M4 relief road around Newport.

Exam practice

Study the OS map below:

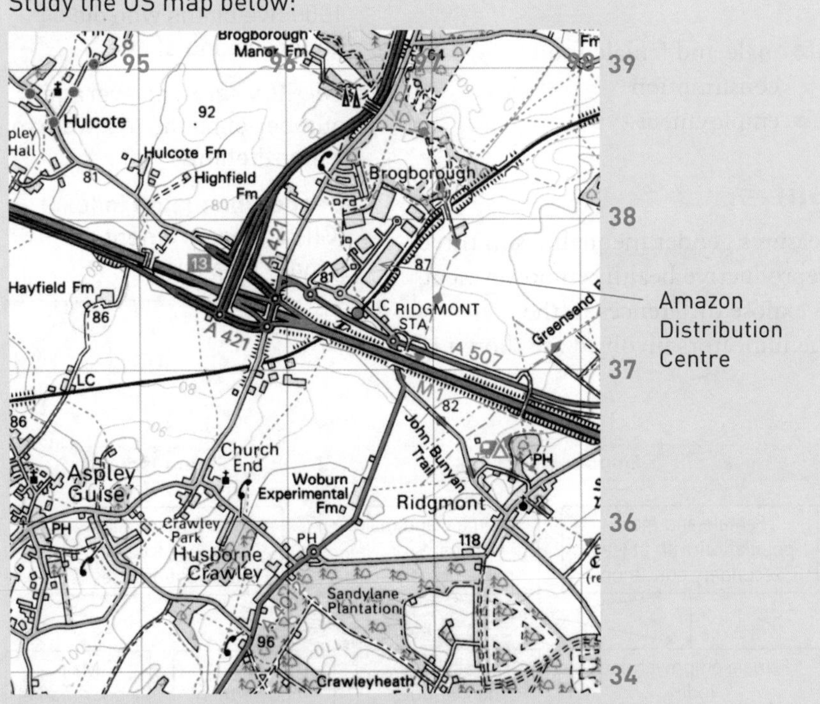

Figure 19 OS map showing location of the Amazon Distribution Centre. Scale 1:50,000.

1 Give a four-figure grid reference for the location of the Amazon Distribution Centre. [1]

2 Give two reasons to explain why this site was chosen by Amazon for a distribution centre. [4]

Exam tip

Maps are an essential tool for geographers. It is certain that in your GCSE examination there will be at least one question which will contain an OS map and it is likely that other questions will contain simple maps and sketch maps. It is therefore essential that you have basic map-reading skills and can:

- read symbols, give grid references and direction, measure distance, use scales and understand contour lines
- describe geographical features shown on a map.

Theme 7 Social Development Issues

Measuring social development

How is social development measured?

REVISED

When considering how to measure development within a country, we often use an economic indicator. As this does not always reflect the standard of living that people within a country experience, instead we need to look at **social development**.

Some indicators that could be used to measure social development include:

- **life expectancy**
- **literacy rates**
- the number of people per doctor
- average food consumption
- the number of homeless people
- deaths from unsafe water and sanitation
- the **infant mortality rate**.

Two areas that are frequently used to measure social development are gender equality in a society and the health of its citizens.

Social development A measure of how well a society is changing for the better or how living standards are improving

Life expectancy The average age a person is expected to live to in a population

Literacy rate The percentage of people in a population who can read or write

Infant mortality rate The number of babies per 1000 live births who die under the age of one

Fertility rate The average number of births to a woman in her lifetime

Gender inequality index (GII) A measurement of gender disparity

Gender

Development indicators based around gender measure the progress that a country is making towards equal rights for both men and women. Many HICs still have gender equality issues. Some social indicators that can give an indication of gender equality are:

- male and female literacy rates
- **fertility rate**
- male and female life expectancy
- male and female food consumption
- employment type.

Gender inequality index (GII)

The **gender inequality index (GII)** measures gender inequalities in three key aspects of human development: reproductive health, empowerment and economic status. It is designed to expose differences in the achievements of men and women. The indicators involved are shown in Figure 1.

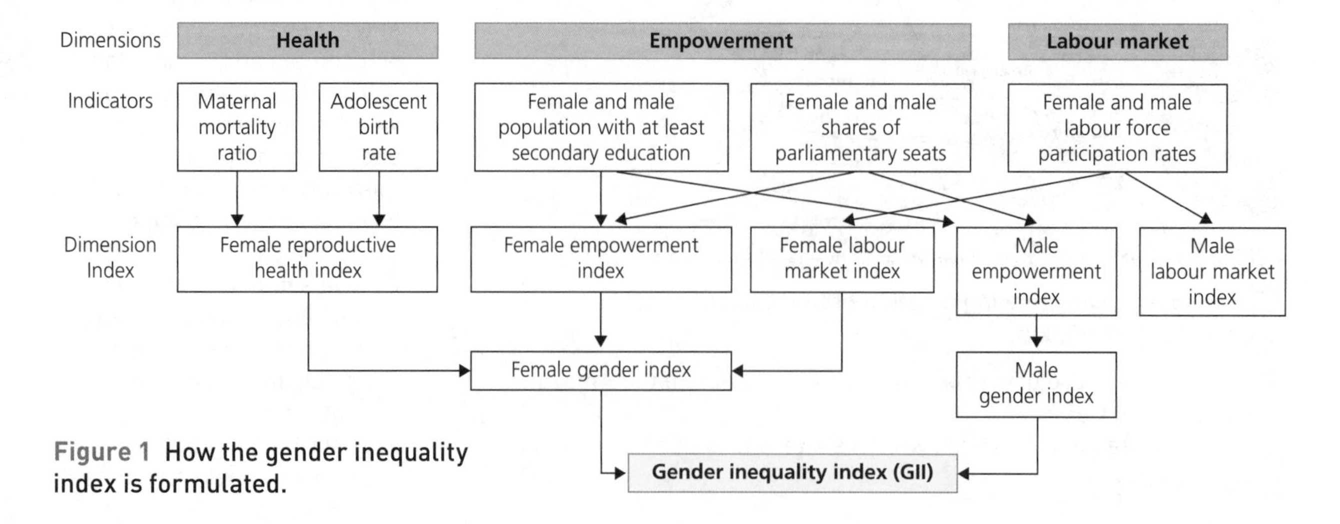

Figure 1 How the gender inequality index is formulated.

Health

Development issues based around health measure the progress that a country is making towards a healthy life for all of its citizens. A range of indicators are used to measure not only the health of the citizens, but also the state of the main health service provided within that country. Such indicators include:

- average life expectancy
- infant mortality rate
- percentage of **gross domestic product (GDP)** spent on healthcare
- length of hospital waiting lists and waiting times
- mortality rate of specific health conditions such as cancer and heart disease.

Continuum of social development

Many of these indicators are interdependent. For example, the length of hospital waiting lists may very well depend on the percentage of GDP spent on healthcare, which will over time influence the average life expectancy. It is too simplistic to describe a country as either 'healthy' or unhealthy' as there are many **variables** involved. Instead, we need to think of a gradual progression or **continuum of social development**. For example, in 2015 the country with the longest life expectancy was Japan with an average life expectancy of 83.7 years, and the shortest Sierra Leone with 50.1 years. Between these two countries there are 181 other countries with different life expectancies. The continuum enables us to appreciate the range of average life expectancies between the longest and shortest.

This continuum is not static as life expectancy changes annually in most countries due to further medical developments or investments. As such, it is useful to look at the **development gap** between countries as a dynamic picture (always changing). As vaccination programmes are put into place in poorer countries, death rates can be cut relatively quickly and therefore the gap between them and richer countries is reduced.

Human development index (HDI)

The **human development index (HDI)** is calculated from four development indicators and measures a country's progress across a range of factors:

- average length of schooling in years
- literacy rates
- **gross national income (GNI) per capita**
- life expectancy.

The HDI brings together social and economic factors and therefore may be a more reliable indicator of measuring overall development.

Now test yourself

TESTED

1. What do you understand by the term 'social development'?
2. Take two examples of gender indicators and two examples of health indicators and explain how these inform us as to how well developed a country is.
3. Explain why the HDI is a useful measure in some countries.

Gross domestic product (GDP) The total value of goods and services produced in a country in a year

Variables Factors that can change and influence an outcome

Continuum of social development A way of thinking about social development as a continuous process that does not have an end point

Development gap The gap that exists in the measurement of development between the world's richest and poorest countries

Human development index (HDI) A measure of the development in a country taking into account wealth, education and average life expectancy

Gross national income (GNI) per capita The average income in a country per person

Exam tip

When asked about gender indicators or health indicators, you should include a number of examples of each and explain how these measures are beneficial.

Exam practice

1. Adult literacy rate is a social development indicator. Describe what is being measured by this indicator. [2]
2. Explain why people use the term 'continuum of social development'. [4]
3. Evaluate the advantages and disadvantages of using the HDI to measure social development. [8]

Uneven social development

What challenges face social development in sub-Saharan Africa and South Asia?

REVISED

Birth and death rates

Population growth depends on the balance between **birth rates** and **death rates**. Social (S), economic (E) and political (P) factors influence these rates.

Birth rate The number of births per 1000 people per year

Death rate The number of deaths per 1000 people per year

Population pyramid A graph that shows the age and gender distribution of a population

Factors that lead to higher birth rates	Factors that lead to lower birth rates
Children provide labour on farms and security for old age (E)	People tend to marry later and therefore have reduced child-bearing years (S)
Large families are seen as a sign of virility (S)	Women are educated and often follow careers which delay or prevent them starting families (P)
Girls may marry early and therefore extend their child-bearing years (S)	The high cost of living means it is expensive to raise children (E)
Women may lack education and stay at home to raise a family rather than work (S)	Couples prefer to spend money on material things such as holidays and cars (E)
A high infant mortality rate encourages larger families to ensure survival of some children (S)	Birth control is readily available (P)

Factors that lead to higher death rates	Factors that lead to lower death rates
HIV, Ebola and other difficult to control diseases are having an impact on death rates in LICs (S)	Better healthcare and vaccination programmes are more available to people (P)
In HICs, the increasingly higher proportion of elderly people in ageing societies is leading to an increase in death rates (S)	Less physically demanding jobs put less stress on people physically (S)
	People are educated about health and hygiene (P)
	Water supplies are more reliable and cleaner (P)
	There is more sanitary disposal of waste (P)

Population pyramids

Which of the above factors has the greatest influence on a country's population will vary but we can see how the structure of the population is altered by looking at **population pyramids**. These graphs divide the population into five-year age groups which are shown as horizontal bars and then the graph is split into two to show males and females. Two examples of population pyramids are shown in the following table, together with factors which may affect their structure (the data is representative of 2014).

Sub-Saharan Africa: Nigeria (birth rate 38 and death rate 13)	South Asia: India (birth rate 19.8 and death rate 7.3)
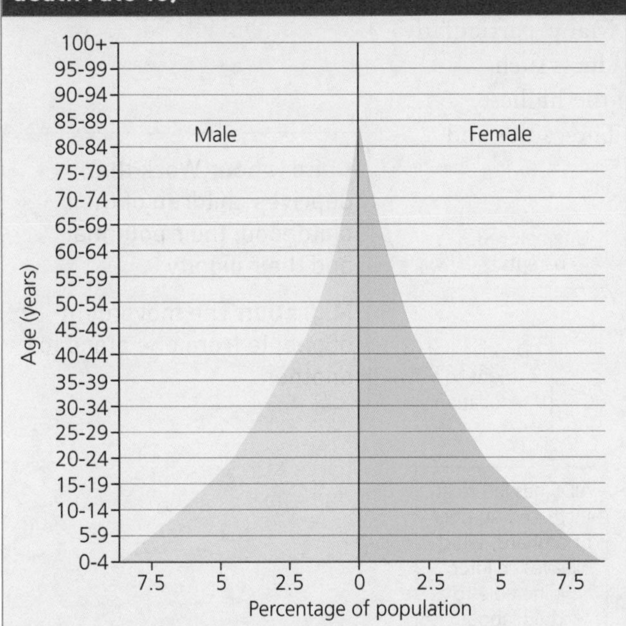	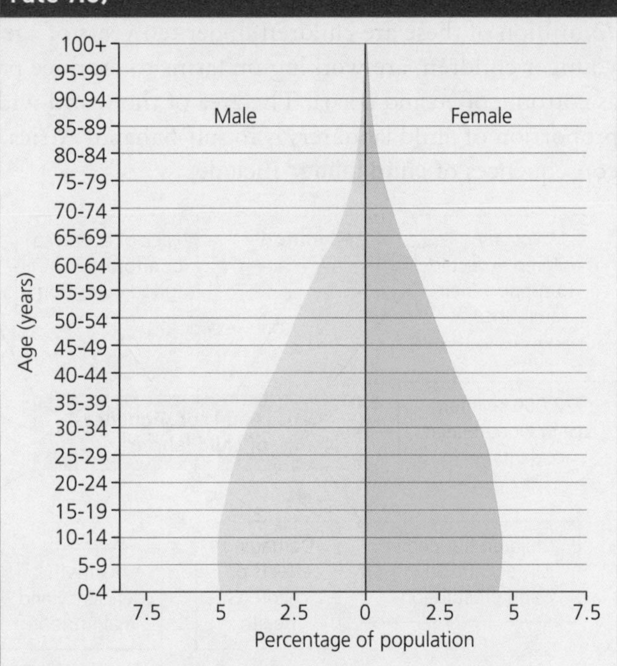
A high infant mortality rate of 74 deaths/1000 live births Average life expectancy of 52 years Large families: fertility rate 5.25 Low use of contraception In 2012, 3 per cent of the population had HIV/AIDS Poor sanitation and drinking water A high risk of catching a major infectious disease Literacy rates in 2010 were only 61.3 per cent 0.4 doctors per 1000 population Civil unrest since 2002 by Boko Haram The Nigerian government has a large amount of foreign debt In 2014, Nigeria ranked 181 out of 191 countries in the world assessed for its political stability – a high chance of civil, unrest, terrorism or a coup Expenditure on health is 5.3 per cent of GDP (2011)	An infant mortality rate of 43 deaths/1000 live births Average life expectancy of 67.8 years Family size decreasing: fertility rate 2.5 Medium use of contraception In 2012, 0.3 per cent of the population had HIV/AIDS Poor sanitation and drinking water A high risk of catching a major infectious disease Literacy rates in 2010 were only 62.8 per cent 0.65 doctors per 1000 population In 2015, India was considered to have the seventh largest economy in the world but it continues to face issues of poverty and corruption In 2014, India ranked 165 out of 191 countries in the world assessed for its political stability – a medium chance of civil, unrest, terrorism or a coup Expenditure on health is 3.9 per cent of GDP (2011)

Revision activity

1. Draw the outline of the two population pyramids. Add labels to show key characteristics of each pyramid.
2. For each country, draw a table similar to the one below to show the social, economic and political factors that have created the current population structure

Social factors	Economic factors	Political factors

Exam tip

The specification asks you to look at an example of a South Asian and a sub-Saharan African country, therefore you must learn a detailed example of each.

Reasons for child labour

It is estimated that there are currently 168 million child workers and 73 million of these are children under ten years of age. Many, particularly younger children, are working on farms to produce products such as cotton, coffee and cocoa. The area of the world with the highest proportion of child labourers is in sub-Saharan Africa. The causes and consequences of **child labour** include:

Child labour Work that deprives children of their childhood, their potential and their dignity

Migration The movement of people from one place to another

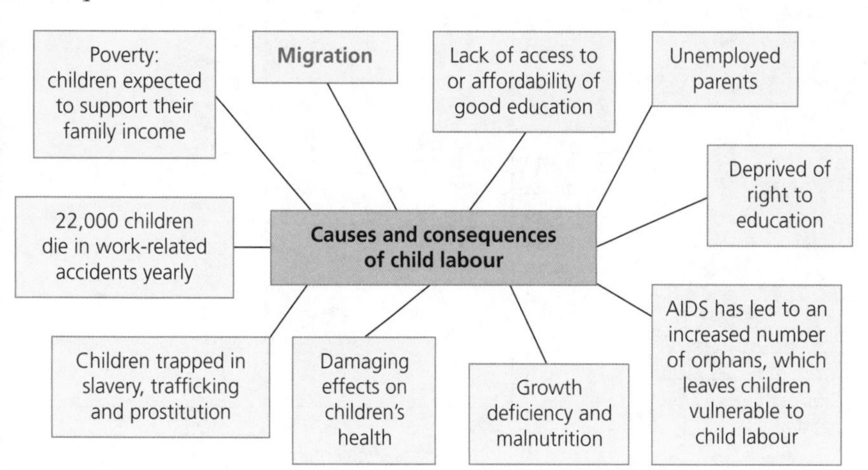

International organisations tackling child labour

The difficulty in trying to tackle child labour is due to the range of countries involved and the varying factors that cause the child labour at each location. There are a range of organisations working internationally to try to end child labour. These include:

- The International Labour Organisation (ILO). It collects data from different countries and uses this data to set targets which can be used as a benchmark to monitor progress. The ILO then makes recommendations to individual governments as to how this can be achieved in their country, which frequently include:
  - Improving access to education for all children so that they can progress and succeed in life.
  - Creating more trade unions so that they can prevent and protect against child labour.
  - Improving social security systems so that the poorest in society are supported rather than them relying on their children.
  - Increasing public awareness of the issue and changing people's attitudes.
- The United Nations (UN) has passed a number of conventions which aim to get international agreement on the issue of child labour. These include Convention 138 on the Minimum Age for Admission to Employment, and Convention 182 on the Worst Forms of Child Labour.
- International World Day Against Child Labour attempts to raise the profile of various aspects involved in child labour.
- Charities such as Child Hope and SOS Children's Villages try to raise awareness as well as working with the communities involved.

Now test yourself

TESTED

1. What do you understand by the term 'child labour'? Can you give some examples?
2. In your opinion, what are the top three causes of child labour? Can you explain why each one leads to an increased number of children in work?

Challenges in primary education

The UN Millennium Development Goal 2 set a target to:

- increase primary school enrolment from 83 per cent in 2000 to 91 per cent in 2015 for developing countries
- halve the number of children not attending school globally.

Although there has been progress, this is not always evenly spread across the whole population in some countries.

Example of challenges in reducing child labour and extending primary education in India

In India, the issue of child labour remains a significant, if decreasing issue. In 2010, there were 4.98 million children in child labour, whereas by 2011 there were 4.35 million child labourers. In addition to poverty, the lack of education is a key cause of child labour. More men than women are educated and of the 62 per cent of children who do not attend school, 62 per cent were girls. Girls often spend less time at school than boys. The reasons for this include:

- Poor quality of school buildings, facilities and teaching in many schools. For other children, schools are too far away or simply too expensive for their family to send them to.
- Attitude towards women in society: many families that follow the **caste system** still have an oppressive attitude towards women and do not see the value in their education.
- Many girls are expected to marry young through arranged marriages.
- The fear that sexual harassment of girls may bring dishonour to the girl's family.

The consequences of this lack of education for girls include:

- poor prospects of being able to live independently
- a higher rate of infant mortality to uneducated mothers
- larger family sizes which keep women in the home and rearing children.

Strategies to improve access to education of girls and to reduce the child labour that poor education creates include:

- Empowering communities: charities working with rural communities and educating parents to help them see the value of a girl's education.
- Locating new schools in places where all pupils can reach them and providing the infrastructure for pupils to get to them.
- The establishment of Bal Sabhas (Girls' Councils) in all primary schools to give girls a voice within the school.
- The Child and Adolescent Labour (Prohibition and Regulation) Act of 1986 makes it a criminal offence to employ children in 64 industries which have been deemed hazardous. Therefore with the loss of the 'value' of putting children to work, it is hoped that families will see the value in their education.
- The Right of Children to Free and Compulsory Education Act of 2009: this Act aims to make free and compulsory education widely available to children up to the age of fourteen.
- Local initiatives working with local leaders and trade unions to create child labour-free working environments.

Revision activity

Draw a mind map to summarise all your information on issues in primary education. Begin with 'primary education' in the middle of the page and then add three arms: one each for causes, consequences and solutions. Remember to include case study information.

Caste system An Indian class system which involves determining social class by the one you were born into

Reasons for international refugees and asylum seekers

There are many different reasons why people migrate. Some people are **economic migrants** and choose to move (due to **pull factors**), whereas other people are forced to move (due to **push factors**) and are refugees or **asylum seekers**. Refugees are often **international refugees** as they are fleeing from conflict, persecution or a natural disaster in their own country (push factors) and seek a safer life in another country.

Since 2000, there has been an increasing number of refugees trying to enter Europe from countries in Africa, the Middle East and South Asia, mainly from Syria, Afghanistan, Iraq, Eritrea and Somalia. This movement of people is the largest since the Second World War. With such large numbers of people being displaced and travelling large distances, the impacts are felt in many countries. For example, Lebanon is a country of 4.4 million inhabitants and has been greatly affected by the refugee crisis. By May 2016, the country was hosting around 1.1 million Syrian refugees (both registered and non-registered), 42,000 Palestinian refugees from Syria, 6000 Iraqi refugees and nearly 450,000 refugees from Palestine.

Economic migrants People who move with the hope of earning more money elsewhere

Pull factors Factors that attract people to a place

Push factors Factors that make people want to leave a place

Asylum seekers People who have applied for legal recognition as refugees in another country and are waiting for a decision

International refugees People who are forced to leave where they live to move to another country

Example of a country dealing with the impacts of refugees: Lebanon

Lebanon is a small country of 4.4 million people which borders the Mediterranean Sea, Israel and Syria. Due to its location it receives many migrants fleeing from the Syrian civil war and other countries, as we have seen above. Some of the impacts this is having on the country are:

- The population has grown by 25 per cent.
- It has the highest per capita concentration of refugees in the world.
- Increased pressure on infrastructure, public health, labour, education, housing and security.
- The Lebanese government has asked the international community for $449 million in assistance to host the refugees.
- Security cells have been established in local communities to record violent incidents, illegal business and so on.
- Tent cities and squatter settlements cover large areas in Lebanon.
- Limited access to clean water and sanitation lead to spread of disease.
- Many refugee children miss out on schooling.

Tackling the issue of refugees

National governments in Europe have reacted to the mass migration of refugees in a variety of ways in an attempt to manage the influx of people:

- Germany and Sweden see the refugees as victims and have welcomed them to their countries and help them to integrate into their societies.
- Austria is trying to limit the number of refugees to 80 a day.
- The UK has agreed to accept 20,000 refugees from Syria by 2020 and it will accept more unaccompanied Syrian child refugees.

Now test yourself

1. Explain the difference between a refugee and an economic migrant.
2. Group the causes of the European migrant crisis into social, political and economic causes.
3. Explain why the refugee crisis is particularly intense in Lebanon.

TESTED

International agreements

In Europe there are international agreements in place with regard to the movement of people across countries. One of these agreements is known as the **Schengen agreement**, which was signed in 1995. This created Europe's Schengen Zone which enables passport-free movement of people between the countries that have signed the agreement. In 2016, there were 26 Schengen countries: 22 European Union (EU) members and four non-EU (Norway, Iceland, Switzerland and Liechtenstein). There are therefore six EU countries which have not signed up to this agreement and prefer to manage their own borders (Bulgaria, Croatia, Cyprus, Ireland, Romania and the UK). This itself has created issues. For example, the restriction of free movement into the UK has led to an increase in the number of illegal migrants trying to get into the country. One route used by illegal migrants was to get to Calais and attempt to cross the English Channel illegally by stowing away in lorries or vans, or by attempting to get through the Eurotunnel. This led to a slum area developing in Calais known as 'the jungle', where migrants lived while attempting to find a route to the UK. The jungle was demolished in 2016 in an attempt to reduce the number of illegal immigrants coming to the UK.

Schengen agreement An EU agreement whereby border checks between some member states have largely been removed

Revision activity

Draw a bubble diagram to show the responses to the European refugee crisis. Shade in the bubbles that are national responses in one colour and those that are international responses in another colour.

However, with the increasing numbers of migrants from Africa and Asia reaching Europe illegally (for example by crossing the Mediterranean Sea in inflatable boats) and hence avoiding border controls, this has led to the following changes:

- In 2016, border controls were temporarily introduced in seven Schengen countries (Austria, Denmark, France, Germany, Norway, Poland and Sweden).
- An EU naval operation, Operation Sophia, has been put in place to monitor the Mediterranean Sea to prevent human smuggling and trafficking.
- EU member states agreed to provide task forces of national experts and support teams to work in hotspots such as Greece and Italy to expedite refugee screening.

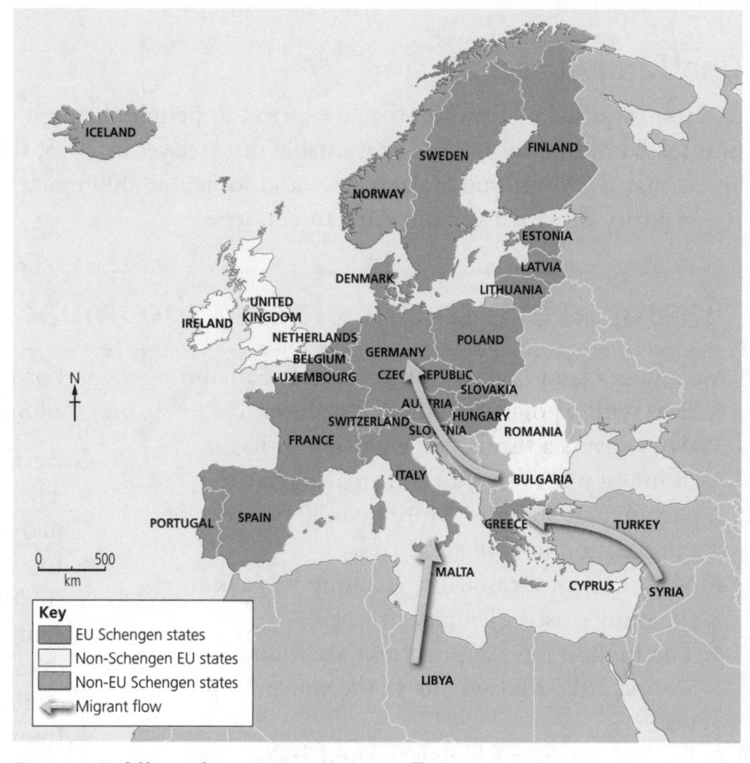

Figure 2 Migration routes across Europe.

Now test yourself

TESTED

1. Give two contrasting ways in which countries have attempted to deal with the refugee crisis in Europe.
2. For each of the methods above, describe the advantages and disadvantages of this approach.

Exam practice

'Eradicating child labour is a key factor in the further development of countries in both South Asia and sub-Saharan Africa.' To what extent do you think this is true? [8]

Exam tip

If an exam question asks you for the consequences of child labour or lack of primary education then remember to link the effect to the issue. You have to not only state what the consequence is but also explain how that results because of the issue.

What are the healthcare issues in sub-Saharan Africa?

REVISED

Reasons for high infant mortality rates

Sub-Saharan Africa suffers from the highest infant mortality rates in the world. In 2015, there were 86 deaths per 1000 live births, and this rate is even higher for newborn babies. Although the region is the poorest area in the world and cannot afford a good level of medical care, other factors include:

- Neonatal infections: a high rate of infection from the process of delivering the baby causes a high rate of infection in newborn babies.
- Around ten per cent of deaths in early childhood in the region are due to diarrhoea.
- The lack of skilled birth attendants leads to many children dying within 24 hours of being born.
- Lack of vaccinations and mosquito nets to stop preventable diseases.

Two of the most common killers in sub-Saharan Africa are **malaria** and HIV.

Malaria A serious tropical disease which if left untreated can be fatal; symptoms include fever, headaches, vomiting and muscle pain

Challenges of malaria

Malaria is caused by parasites that are spread to people through the bites of infected mosquitoes. It is a preventable disease, yet in 2015, there were an estimated 214 million malaria cases and some 438,000 malaria deaths, the majority of which were in African countries.

Example of a country facing the impact of malaria: Malawi

Malawi is a land-locked country in sub-Saharan Africa, with a population of 16.8 million. Lake Malawi covers a third of its land area. It has a high infant mortality rate and an average life expectancy of 50 years. More than 80 per cent of the population lives in rural areas.

- Malaria varies seasonally, reaching its peak in the rainy season (January to April).
- The highest rates of infections are found around Lake Malawi due to the warm, stagnant water.
- Infection rates are higher in rural areas.
- Children, pregnant women and those with HIV are at particularly high risk.
- Mosquitoes are becoming resistant to insecticides.
- For most people in Malawi, visiting a doctor is a long walk away.

Government strategies to combat malaria:

- The Malaria strategic plan sets targets and monitors intervention.
- Increase the use of insecticide-treated bed nets (ITNs) which cost around £3. In 2015, 80 per cent of households had at least one net, but this does not mean that all members of the family could sleep under it.
- Improve access to fast and effective treatment so that early symptoms can be managed.
- Indoor residual spraying (IRS): this involves spraying insecticides in places where mosquitoes are more likely to come into contact with people.

Now test yourself

TESTED

1. Describe the strategies that have been used in the attempt to control malaria.
2. Explain why malaria is difficult to control.

Exam tip

If a question asks you to explain the challenges that malaria presents then you must be able to link the challenge back to the country or location it is in and why the conditions there make it a challenge.

Challenges of HIV

There were approximately 36.9 million people in the world living with **HIV/AIDS** by the end of 2014, and 2.6 million of these were children. Sub-Saharan Africa has the most widespread HIV/AIDS epidemic in the world. In 2013, an estimated 24.7 million people were living with HIV, which accounts for 71 per cent of the global total.

Example of a country facing the impact of HIV/AIDS: Malawi

Malawi has an estimated 1 million people who are infected with HIV.

- Average life expectancy is 50 years, largely due to deaths from AIDS.
- Rates of HIV infection are higher in urban areas than in rural areas.
- Many families are in poverty due to adults being too ill to work.
- The country's development is limited due to a reduction in taxes paid as a result of fewer people working.
- Children of adults with HIV often drop out of school to care for their parents.

Government strategies to combat HIV/AIDS:

- HIV testing and counselling (HTC) services have increased over the past few years and during 2012–13 over 2.1 million HIV tests were carried out.
- Large investments in preventing mother-to-child transmissions where pregnant women are given access to medication while pregnant, which helps to prevent the infection being given to the baby.
- An increase in the availability of free condoms.
- An increase in the number of people treated with anti-retroviral treatment (ART), which helps to prevent HIV leading to AIDS and therefore prevents early death.

International response to malaria and HIV/AIDS

Most nations are fully committed to reducing infection and death rates from malaria. The 'Roll Back Malaria' initiative had over 500 partners working together to provide a co-ordinated global response to the disease, and one of the UN's Millennium Development Goals is that the incidence of the disease should have reduced by 2015.

The initial global response to HIV focused on prevention through encouraging behaviour change and also research into a vaccine. However, it soon became evident that this would not be enough to stop the epidemic. Today, the UN AIDS Fast Track Strategy is aiming to end the epidemic by 2030 through improvements in the availability of contraception, education and the availability of medication which prevents the virus leading to AIDS.

HIV Human immunodeficiency virus is a virus that attacks the body's immune system and weakens its ability to fight infections. If left untreated, HIV may lead to AIDS

AIDS Acquired immunodeficiency syndrome is the final stage of HIV infection, which may lead to death unless treated

Now test yourself

TESTED

1 What impact does a high level of HIV infection have on a country?
2 List the different approaches that can be taken to limit HIV infection.
3 Are government-based or international-based approaches more effective at dealing with HIV/AIDS?

Revision activity

Draw a table to compare the impacts of and strategies to tackle malaria and HIV infection.

Exam practice

1 What is meant by the term infant mortality rate? [2]
2 Why is this an important measure of development? [4]
3 Explain why either malaria or HIV is difficult to control and manage in sub-Saharan Africa. [6]
4 Describe the international responses to either malaria or HIV. [4]

Top-down and bottom-up approaches to development

Top-down approach	Bottom-up approach
Decisions are made at governmental level and usually involve a high cost. Communities likely to be affected by the decisions have little say as to what is done	Decisions are made by the local communities that they will affect. They try to help communities by helping them to help themselves
The advantages of these types of schemes are that they may be part of a strategic plan which aims to develop the infrastructure of the country. However, they frequently lead the country into debt and the jobs that are created are often not for the local community	The advantages of these types of schemes are that they are small scale and so cost less, are more sustainable and usually meet the needs of the local community better

Top-down approach
Large-scale projects that are decided on by national governments

Bottom-up approach
Projects that are planned and led by local communities to help their local area

Example: Kaste Dam, Lesotho

The Kaste Dam is part of the Lesotho Highlands Water Project. This project was developed through a partnership between the South African and Lesotho governments to improve the water supply for South Africa and provide Lesotho with an income. There are many health benefits that this improved water supply has brought although the environmental social benefits, as far as the farmers who lost their land are concerned, are difficult to see.

Figure 3 Kaste Dam in Lesotho.

Example: WaterAid

This British charity is helping to put hand water pumps into Ethiopian villages. In Ethiopia, 42 million people do not have access to safe, clean drinking water and over 9000 children die each year from diarrhoea caused by dirty water. WaterAid works with each community, providing the hand pump and showing the community how to maintain it so that they do not have to work for hours each day to collect water, which leaves the villagers more time to farm.

Figure 4 Women collecting water from a well in Ethiopia.

Measuring progress of development

There are a number of ways in which the developmental progress of sub-Saharan countries can be measured. Some of these are the Millennium Development Goals which measured progress from 2000 to 2015. These have now been replaced by the Sustainable Development Goals. See the table below for more details.

Method of measurement	Progress
Millennium Development Goals: ● MDG 1: eradicate extreme hunger and poverty ● MDG 2: achieve universal primary education ● MDG 4: reduce child mortality ● MDG 5: reduce maternal mortality	● 8 out of 26 countries made no progress in the last decade ● 5 out of 43 countries made <50 per cent progress towards target ● 27 out of 43 countries made 50 per cent or more progress to achieving this goal ● 18 out of 43 countries made <50 per cent progress towards target
Sustainable Development Goals: 17 goals which aim to end poverty, protect the planet and ensure that all people enjoy peace and prosperity	These were put in place in 2016 and progress will be measured by the United Nations Development Programme (UNDP)
Human development index (HDI): A complex development indicator which takes the following into account: ● life expectancy at birth ● expected years schooling for school-age children ● average years of schooling in the adult population ● gross national income (GNI) per capita	● Botswana in 2014 had a medium HDI of 0.698 ● Angola in 2014 had a low HDI of 0.532 ● Niger in 2014 had a very low HDI of 0.348

Now test yourself

1. What do you understand by the term 'top-down approach?'
2. How does this differ from a 'bottom-up approach'?
3. What factors make measuring progress difficult?

Revision activity

Research one bottom-up approach and one top-down approach to improving health in sub-Saharan Africa. Then draw a table similar to the one below highlighting the key features and the extent to which they are socially, environmentally and economically sustainable.

Scheme – key features	Socially sustainable	Environmentally sustainable	Economically sustainable

Exam practice

'In the twenty-first century, bottom-up approaches rather than top-down approaches are the way forward to speed up development in sub-Saharan Africa.' To what extent do you agree with this statement? [8]

Theme 8 Environmental Challenges

Consumerism and its impact on the environment

What are the impacts of increasing consumer choice on the global environment?

REVISED

What is consumerism?

In the twenty-first century, **consumerism** dominates Western societies more than ever. We all buy goods, many of which are essential for us to survive (food and clothing and so on). However, we also buy many non-essential goods which drives our consumer society even more. For example, globally in 2014 there were 7.2 billion mobile phones in use, and a world population of 7.19 billion. Therefore, we are now using more mobile phones than there are people in the world. Considering that in the 1980s it was uncommon for people to have mobile phones, the growth in popularity of this product is phenomenal. However, most people are not content to own a mobile, but they want the latest model with the most up-to-date functions and services. Therefore, consumerism is constantly being driven by new and improved, non-essential products.

Ecological footprint

Consumerism also drives the food industry. In the 1950s and 1960s, it was only possible to buy fruits like strawberries and apples when they were in season in the UK. At that time the UK produced a much higher proportion of its food than we do today and consumers were more in touch with the growing seasons of the food they ate. However, with increased globalisation we now have seasonal fruits available in our supermarkets throughout the year. These foods are sourced from all over the world, for example:

- lamb from New Zealand
- beef from South America
- apples from Chile
- green beans from Kenya.

All of these foods can be grown or raised in the UK but may not be available all year round and therefore thousands of **food miles** (both air and sea) are covered each year in transporting these foods from their source country to the place they are consumed. This transportation has a large impact on the environment through pollution, and in changing the use of the areas of land that are used to grow the food. These both have an impact on the **ecological footprint** of the way that we consume food. When considering the ecological footprint of goods that we consume, the following factors are taken into consideration:

- the energy used in their formation
- the land area taken to produce the goods
- the amount of carbon produced from production to consumption
- the impact on the ecosystem where produced
- the waste that is created from producing and consuming the product.

Consumerism The idea that it is good if people buy an increasing amount of goods or services

Food miles The distance that food travels from where it is grown or reared to where it is consumed

Ecological footprint A measure of the impact on the natural environment a person's lifestyle has. It is measured as the land area that it takes to sustain this lifestyle

Exam tip

An exam question may ask you to 'explain the links between …'. Remember that links can go both ways and therefore look at both how consumerism affects global interdependence and also how global interdependence affects consumerism.

Now test yourself

1. What do you understand by the term 'consumerism'?
2. Explain why consumerism has an impact on the ecological footprint of a population.

TESTED

Global interdependence, consumerism and ecosystems

The buying and selling of goods between countries is known as **global interdependence**. Countries rely on other countries to supply goods or services which their populations consume. The increase in consumerism increases the impacts that it has on both local and global ecosystems. The impacts of consumerism on the environment depend on the fragility of the **ecosystem**, the scale of the growth/development of the product and how much the source country depends upon the product to contribute towards its income.

Global interdependence When countries depend on each other to buy or sell goods

Ecosystem A community of plants and animals and how they interact with the environment within which they live

Monoculture Large-scale farming of one crop type

Example of impacts on tropical rainforests: Borneo

Borneo is an island in South-East Asia and divided between the countries of Malaysia, Brunei and Indonesia. The island contains a tropical rainforest biome. Large areas of the rainforest are being cleared to create palm oil plantations. The increased demand for palm oil impacts on the tropical rainforests in Borneo.

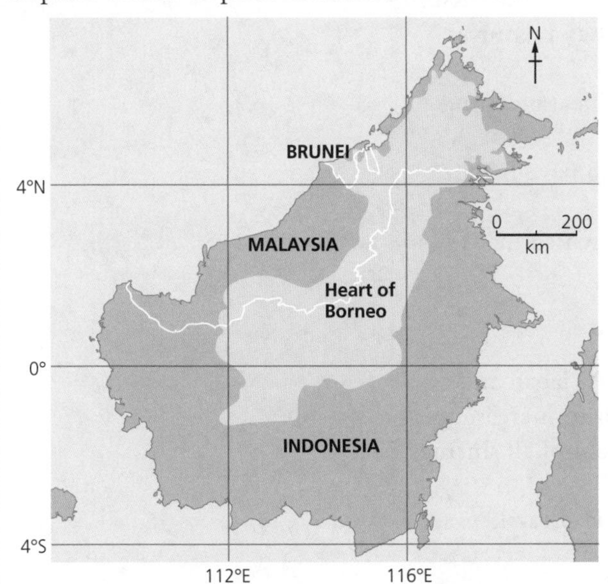

Figure 1 **Borneo and the 'Heart of Borneo'.**

Consumerism

- Palm oil is widely accepted as a healthy alternative to other cooking oils and in an increasingly health-conscious Western world demand has increased.
- It can also be used as a biofuel which provides a greener solution to the energy crisis.
- 66 million tonnes of palm oil are used annually.
- Palm oil is used in many products including biscuits, margarine, make-up and soap.

Ecosystem

- Borneo is thought to contain six per cent of the world's species due to its vast areas of rainforest.
- Palm oil plantations are grown as a **monoculture** and therefore reduce the biodiversity compared to the rainforest they replace.
- Large areas of rainforest are destroyed, not only for the plantation itself but also for the infrastructure that is needed for the plantation.
- Local indigenous people and animals become displaced.
- Air and water pollution, together with soil erosion, frequently result.

Global interdependence

- Palm oil has higher yields and lower production costs compared to many of its alternatives and therefore is a highly profitable produce for the newly industrialised countries (NICs) that the island of Borneo is governed by.
- Indonesia and Malaysia are the largest palm oil producing countries in the world and it forms a major part of their exports.
- Huge profits are made for the multinational corporations that invest in the plantations. In recent years these corporations have been asked to support sustainable use of land for palm oil production.
- The biggest importers of palm oil are India, the European Union and China.

Now test yourself

TESTED

1. What do you understand by the term 'ecological footprint'?
2. What factors will increase a person's ecological footprint?
3. Describe the factors that have led to the destruction of a tropical rainforest that you have studied.
4. Make links between the factors that you have described in question 3.

Example of impacts on mangrove forests: Bangladesh

Bangladesh is a South Asian country to the north of the Bay of Bengal and is the eighth most populated country in the world. Due to its extensive coastline there are large areas of mangrove forests which provide an ideal habitat for shrimps (prawns). The increased demand for shrimps impacts on the mangrove forests in Bangladesh.

Consumerism

- In 2010, over 3 million tonnes of wild shrimp were caught by fishermen in mangrove forests.
- Fish (including shrimp) are the second largest export for Bangladesh at a value of $569.9 million in 2016.
- Large businesses cut down areas of mangrove forests so that the area can be developed for **aquaculture**.

Aquaculture
The commercial farming of fish and shellfish

Ecosystem

- Trees in mangrove forests can tolerate both salt and fresh water and therefore are a valuable habitat for a wide range of animals and fish.
- The roots of the trees hold together the mud and act as a natural defence to coastal flooding.
- The sheltered water created by the trees creates the ideal breeding ground for fish and shrimps.
- 25 million hectares of mangrove forest in Bangladesh have been destroyed, largely to make way for shrimp farms.
- Shrimp farms produce organic waste and chemicals which may pollute natural water sources.

Global interdependence

- Due to high demand from countries such as the USA, Japan and Western Europe, shrimp farming began in the 1970s to increase the supply of shrimps, which now accounts for 55 per cent of all shrimps produced.
- Most shrimp farms are in China, Thailand, Indonesia, Brazil, Ecuador and Bangladesh.

Figure 2 A mangrove forest in Bangladesh.

Now test yourself

1. Describe the factors that have led to the destruction of an ecosystem other than tropical rainforests that you have studied.
2. Explain the role that consumerism has had in this destruction.

TESTED

Impacts of consumerism on the environment

The volume of goods that we buy impacts on both the global and local environments where these are consumed.

Agri-business:
- Large companies own multiple farms growing a single crop
- Chemical fertilisers are used and hedgerows removed to maximise yield and profit
- Habitats are destroyed or polluted
- Little investment goes back into the farms

Transport:
- Complex network of air, sea and land transport moves goods globally
- Season-free availability of foods
- Negative impacts include food miles and water pollution

Impacts of consumerism on the environment

Disposal of waste:
- In 2010, the total amount of household waste in the UK was 27 million tonnes
- Despite government targets, by 2014 the amount of waste was virtually the same, although more is now recycled
- E-waste comes from electrical devices and usually contains metal, which should be recycled but is often shipped to NICs or LICs

Revision activity

1. Draw a bubble diagram to show the factors you think need to be taken into consideration when calculating an area's ecological footprint.
2. For the tropical rainforest example that you have studied, put the factors that have caused their destruction into a Venn diagram similar to the one below. (Make sure that if there is a link between one factor and another, you put them in an overlapping part of the diagram.)
3. Repeat this activity for the second biome that you have studied.

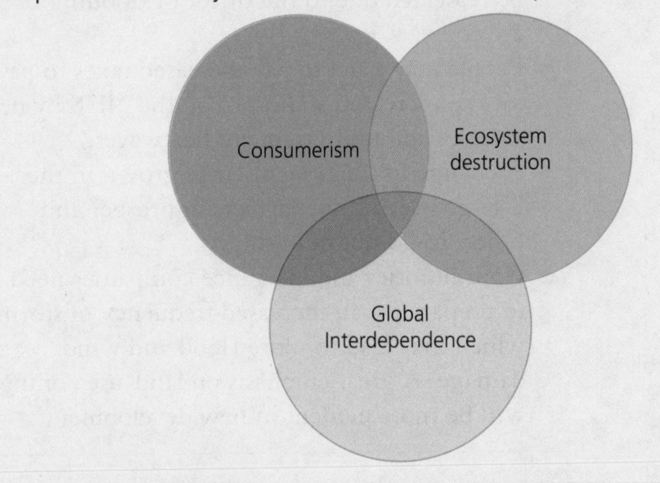

Exam practice

1. What do you understand by the term 'consumerism'? [1]
2. Explain why one biome that you have studied is being destroyed due to consumerism. [4]
3. 'Destruction of ecosystems is necessary if global interdependence is to continue.' How far do you agree with this statement? [8]

How might climate change affect people and the environment?

REVISED

The growth, manufacture and transportation of food and goods around the world increase greenhouse gas emissions causing climate change. The effects of this change can be both long and short term on:

- People's lifestyles and way of life: human activities may have to be adapted to cope with climate change (for example flooding) in addition to limiting the further release of greenhouse gases.
- The economy: the costs of dealing with the effects of climate change and adapting lifestyles.
- The environment: ecosystems may be endangered or even destroyed.

Example of effects of climate change on the UK

People

- Currently it is estimated that about 330,000 properties are at risk of flooding, which could increase to between 630,000 and 1.2 million by the 2080s.
- Whole village communities may need to be abandoned due to increased flooding and/or coastal erosion.
- Increased pressure on the health service to deal with the effects of a heatwave.
- Through drier summers, 27–59 million people may be living in regions affected by water supply deficits by the 2050s.
- Spread of new diseases through new species now being able to survive the warmer UK climate, for example mosquitoes spreading malaria.

Economy

- Drier summers may increase income through tourism.
- The economic cost of flood damage (repairs to buildings, roads, lost days' work, replacing valuables and so on) is predicted to increase to around £27 billion by 2080.
- Insurance premiums will rise as more claims are made. Some areas may become uninsurable.
- Increased food price volatility may cause an increase in the cost of food in the UK.
- New crops such as oranges could be grown in the UK, which will reduce imports.

Environment

- Milder and wetter winter weather.
- A higher frequency of storms, which may also be more severe.
- Increased risk of flooding, particularly in the south-east of England.
- Extremely wet winters are five times more likely over the next 100 years, which may lead to an increased risk of flooding, particularly in the south-east of England.
- Warmer drier summers are likely, which could bring an increased risk of droughts and heatwaves.
- Species of animals and plants may migrate north as they no longer fit the habitat that they currently exist in. Some species may become extinct. New species which previously were unable to exist here may spread into the UK.

Alternative geographical future

- Many coastal communities would need to be relocated due to the threat of flooding or erosion.
- People may need to pay increased taxes to pay for the increased strain put on the NHS by new diseases and more frequent heatwaves.
- A change in the type of crops grown in the UK may lead some farmers to prosper and others to go out of business.
- Communities and insurance companies need to prepare for an increased frequency of storms, which may lead to more flood and wind damage. A great emphasis on land-use zoning will be more evident in new developments.

Now test yourself

TESTED

1. Look at the impacts of climate change on the UK above. Group these into long- and short-term effects.
2. Are there any positive effects of climate change?
3. Explain what you think the geographical future of the UK will be with effects of climate change.

Example of effects of climate change on Tuvalu

Tuvalu is a collection of nine islands that are located in the south Pacific Ocean, off the north-east coast of Australia. The highest point in the islands is only 4.5 m above sea level, but most of Tuvalu is below sea level.

People

- Reduction in food supply due to salinisation of the soil.
- Increased water supply scarcity due to contamination from saltwater and irregularity of rainfall.
- Increase in the number of tropical storms destroys what little resources the islanders have.
- Increase in water-borne diseases which threaten lives.
- Some islanders have already decided to leave and move to New Zealand, becoming environmental refugees.

Economy

- The economy is based on the export of copra (dried coconut kernel used to extract coconut oil) and the sale of fishing licences. These are under threat due to flooding and warmer seas.
- The country has sold its internet domain '.tv', which has guaranteed an income of $50 million over twelve years. This money is being used to help pay for flood defences.

Environment

- The warming of the ocean around Tuvalu decreases the biodiversity on the delicate coral reefs and therefore restricts a food source.
- Increased amounts of stagnant water due to frequent flooding.

The soil of Tuvalu is prone to increasing salinisation due to sea level rise, which threatens the habitats of some plants such as coconut trees and pulaka.

Alternative geographical future

- The islanders will need to adapt to a life living below sea level, relying completely on robust sea defences to prevent flooding
- A large proportion of the population may wish to migrate to begin a new life in a less threatened environment.
- The loss of identity of the population as they become settled in other countries.

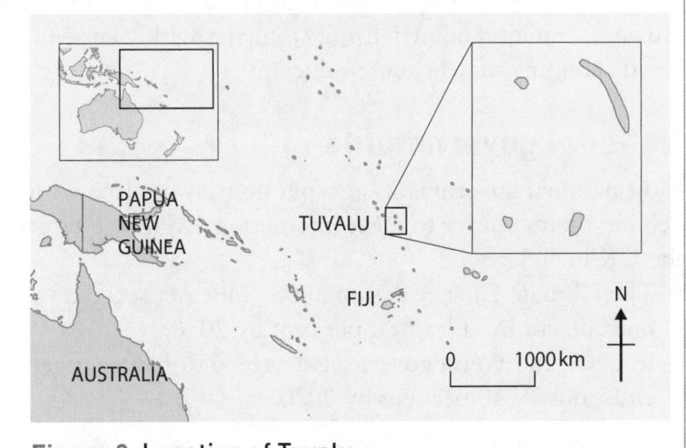

Figure 3 Location of Tuvalu.

Now test yourself

TESTED

1. Identify the long- and short-term impacts of climate change on Tuvalu.
2. What do you think the future for Tuvalu will be?

Exam tip

You must learn the impacts of climate change in two contrasting environments. The UK must be one and your teacher will choose another. Make sure you know the differences in impacts of climate change between these two environments and why the impacts differ.

How can technology be used and people's lifestyles changed to reduce these impacts?

REVISED

Now that it is widely accepted that people are contributing to climate change, if we alter our behaviours then we may be able to limit its impacts. The causes of climate change can be tackled by governments at a global, national and local scale, as well as at an individual level.

International agreements

There have been several attempts to draw up international agreements to restrict emissions from countries. The main ones are:

- The Kyoto Protocol, which was signed in 1997 and committed countries to targets for reducing greenhouse gas emissions between 2008 and 2012.
- The Paris Agreement, which was signed in 2015 and recognised that countries must keep global temperature increases to below 2 °C above pre-industrial levels. It also recognised that newly industrialised countries (NICs) need to cut emissions at different rates from high-income countries (HICs).

International agreements are made increasingly difficult by the different values and attitudes held by various countries. For example, a country in sub-Saharan Africa which suffers increased and prolonged periods of drought may be very willing to sign up to agreements on climate change. However, rapidly industrialising countries with a large manufacturing-based economy may be more reluctant.

National governments

Most national governments also put policies in place with the aim of helping their country to meet internationally agreed targets. Examples for the UK include:

- The Climate Change Act of 2008. This Act sets out that UK emissions must be cut by at least 80 per cent by 2050.
- In 2009, the Welsh government agreed to reduce greenhouse gas emissions by 40 per cent by 2020.

To help achieve these targets the government has invested in new technologies such as:

- Low carbon energy sources: investment in alternative sources of energy such as solar, wind, hydroelectric power and tidal.
- Carbon capture: a method of capturing the carbon dioxide that is emitted from burning fossil fuels.
- Increased efficiency in new buildings: so that less heating or cooling will be required.
- Car fuel standards: the development of electric cars together with energy-saving technology that turns engines off when stationary, both of which reduce fuel consumption.

Local government

Local governments have their own strategies to meet targets. This is the example for Swansea:

Example: sustainable energy action plan for Swansea

- Produce energy plans for new schools in Swansea.
- Reduce the carbon footprint at key development sites in Swansea.
- Develop planning guidance for energy efficiency and renewable energy. The £1.3 billion tidal lagoon proposal for Swansea Bay has been backed by a government review and has potential to produce enough electricity for 155,000 homes.
- Plan to generate electricity from wind, biomass and tidal sources.
- Develop a council corporate energy policy and technical design guidance.

Individual actions

The actions that individuals can take to reduce greenhouse emissions include:

- Insulate all windows, doors and lofts in buildings to reduce heat loss.
- Install solar panels to heat water or generate electricity and use energy-efficient appliances.
- Walk, cycle or use public transport.
- Buy locally sourced foods to reduce food miles.

Now test yourself

TESTED

1. Name an international strategy that has been adopted to tackle climate change.
2. What are the successes or failures of this strategy?
3. Why may local strategies be more effective than international strategies?

Revision activity

Copy the table below. Try to include two examples under each type of strategy and then give an advantage and disadvantage of each strategy. (Think about how effective the strategy may be.)

Strategy	Advantages	Disadvantages
International agreement		
National agreement		
Local government		
Individual actions		

Exam practice

1. Give two ways in which agri-businesses impact negatively on the environment. [2]
2. Explain why the disposal of waste impacts on the environment. [4]
3. Describe the long-term impacts of climate change on the UK. [4]
4. How might people's lifestyles change in the future in a warmer world? [4]

Exam tip

You need to understand 'the role of individuals and governments in adopting new technologies and lifestyles to reduce greenhouse gas emissions'. If you are asked a question about 'the role' of a particular group of people, remember you must discuss not only what they do to reduce climate change but also whether this is more or less effective than another group of people.

Management of ecosystems

How can damaged environments and natural habitats be managed and restored?

REVISED

Environmental strategies to manage habitat and biodiversity

Habitats and **biodiversity** are under threat from human activities within the ecosystem, which can lead to a permanent change in the plants and animals which occupy that land area. To combat this and to try to preserve ecosystems, **environmental strategies** are being used to manage a range of habitats.

Example of a managed habitat: tropical rainforest in Borneo

Natural ecosystem

- It is naturally covered in tropical rainforest, a biodiverse ecosystem containing six per cent of the world's wildlife.
- Endangered species include the Sumatran rhinoceros, Borneo pygmy elephant, giant pitcher plants and the orangutan.

Why is the ecosystem being destroyed?

- Large areas of tropical rainforest are being destroyed to be replaced by palm oil plantations, which covered 6 million ha in 2007.
- 56 per cent of natural rainforest on Borneo has been destroyed.

Environmental strategies

- **Debt-for-nature swap**: the USA and Indonesia have agreed a debt-for-nature swap which will divert $28.5 million intended to repay Indonesia's debts to the USA into environmental strategies for improving land-use techniques in the Indonesian part of Borneo.
- The Heart of Borneo project was established in 2007. This established a protected area (similar to a **national park**) of largely untouched rainforest in the middle of Borneo which aims to conserve and maintain the biodiversity of the forest.
- Development of ecotourism within the Heart of Borneo is a main strategy for social and economic development on the island.
- Raising public awareness: Greenpeace and WWF campaign to raise public awareness of the fact that rainforests are being destroyed to make palm oil to go into products such as toothpaste. They encourage the public to buy products with sustainably sourced ingredients.

Habitat A place where plants or animals normally live

Biodiversity The variety of living things

Environmental strategies Methods of managing an area where the primary objective is to care for the environment

Debt-for-nature swap An agreement that poorer nations will spend money on conservation projects and in return richer countries will cancel part of the debt the poor countries owe

National park An area of special countryside which is protected by the state for people to enjoy and to preserve the wildlife

Strategies to restore habitats damaged by people

There are many ecosystems that are vital for the survival of human communities. Tropical rainforests for example provide wood for paper, construction and furniture, tropical reefs provide a vital source of food for local communities, and wetland ecosystems store and filter vital sources of freshwater and help to protect against flooding and erosion. However, when people interact with these ecosystems, inevitably damage can occur which affects the functioning of the ecosystem. It is therefore in people's interests to restore damaged ecosystems.

Now test yourself

1. Identify three ways in which ecosystems can be managed. Give an example of where each of these is being used.
2. For each of these three methods, identify an advantage and a disadvantage of the scheme.

TESTED

Now test yourself and exam practice answers at www.hoddereducation.co.uk/myrevisionnotes

Example of a managed habitat: tropical grasslands (savannah) in Kenya

Natural ecosystem

- Savannah grasslands are found between the tropical rainforest and desert biomes.
- Endangered species include black rhinoceros, African elephant and Grevy's zebra.

Why is the ecosystem being destroyed?

- Kenya is a poor country, and this poverty forces local people into exploiting its natural resources.
- Illegal wildlife trafficking where animals are taken from their wild habitats to be sold illegally to collectors.
- The poaching of wildlife: such as gorillas for meat and elephants for ivory is threatening the survival of these species.
- Human activities break up the natural habitat into small segments. This means that species are confined to areas which are too small to support them or that their migratory routes are disrupted.

Environmental strategies

- The Masai Mara national reserve (similar to a national park) is 1510 km^2 of protected savannah in south-western Kenya. Here, not only is wildlife protected but it also protects the lands that native tribes occupy.
- Encouragement of sustainable tourism to provide an income for local people so that they do not turn to poaching or trafficking animals.
- A debt-for-development swap agreed in 2006 between the Kenyan and Italian governments sets out that €44 million over ten years be paid by the Kenyan government on development schemes (including conservation) rather than be paid back to Italy.
- The Amboseli–Chyulu corridor is a **wildlife corridor** which connects the Amboseli National Park to the Chyulu Hills and allows free movement of animals such as lions, zebras, elephants and giraffes. This corridor was under threat until Disneynature and the African Wildlife Foundation helped to protect 20,000 ha of the corridor.

Example of wetland restoration: China

In China there are 53 million ha of wetland which not only provide a valuable water resource to the vast population, but also help to protect communities against flooding. In 2014, 60 per cent of China's wetlands were in a poor or relatively poor condition, which was impacting on the local communities. Therefore, **wetland restoration** has been a priority, particularly in the Heilongjiang Province, which contains a sixth of the country's wetlands. The key features of this are:

- The Action Plan for Protecting China's Wetland co-ordinates 39 key projects across the country.
- A wetland protection regulation has been put in place in Heilongjiang Province.
- 10,090 ha of trees have been planted.
- 3441 ha of farmland have been converted back to wetland.
- New jobs involved in tourism replace those that have been lost in farming, which helps the sustainability of the plan.

Management of tourism

Each year, the global number of tourists increases due to increasing affluence in HICs and an overall global population increase. There are both costs and benefits to the tourist industry:

- Benefits: increased income to the host country and improvements in infrastructure.
- Costs: both local (overextraction of water and increased demand for food) and global (increased air pollution). The ways in which governments attempt to minimise the effects of tourism vary depending on the ecosystem being affected, the level of development and the value of the tourist industry to the country. Two examples are given on page 144.

Wildlife corridor A strip of habitat that allows wild animals to move from one ecosystem to another

Wetland restoration The process of transforming a wetland area which has been affected by human activity into an area that can sustain a native habitat

Future sustainability of tourism

In a world where we are seeing projected increases in tourism and new locations becoming a focus for visitors, it is important to consider how to make tourism more sustainable. An increase in numbers will benefit an area economically, but environmentally these numbers must be managed sustainably.

- **Responsible travel**: is every journey taken necessary? Do local people benefit from our visit? Do we treat every destination as someone else's home? Do we leave the destination in the same condition in which we found it?
- Ecotourism: does the place we are visiting conserve the environment? Does it sustain the well-being of the local community? Does it educate people on the importance of the ecosystem being visited?
- Ethical tourism: does our visit benefit the people who live at the destination? Does it provide a better income for families in the area? Are the products and services that will be used sourced locally?

Responsible travel Travel where local families benefit economically through jobs and services

Ecotourism Tourism which has a very low environmental impact

Ethical tourists Tourists who consider the needs of the local people and who have a minimum impact on the environment

Example of coral reefs: the Great Barrier Reef in Australia

The Great Barrier Reef is a popular tourist destination with over 2 million visitors each year. Due to the fragile ecosystem, the following management strategies have been put in place:

- The Great Barrier Reef Marine Park was established in 1975 with the purpose of managing the use and protection of the reef as well as managing the communities that depend on the reef for their livelihood.
- Zones have been created within the park where activities are restricted. This ensures that some areas of the reef are untouched by people, allowing the natural ecosystem to recover and biodiversity to increase.
- Most of the tourist boats to the Great Barrier Reef operate under the eco-certification programme which promotes **ecotourism** at such a delicate ecosystem.

Example of deserts: Sharm El-Sheikh, Egypt

Sharm El-Sheikh is situated in the Egyptian part of the Sinai Peninsula. The natural ecosystem is one of desert of sand and rock. The area has been developed into a popular tourist destination, at first for scuba divers but now as a mainstream holiday resort. This has vastly increased the quantity of water that is used in a very freshwater-limited area. The following steps have been taken to manage the resort:

- Drinking water is transferred via pipeline to the city.
- Desalination plants now use seawater to increase the water supply. This reduces overextraction of the limited supply of freshwater to the ecosystem.
- Sharm El-Sheikh is attempting to become the world's prime destination for ecotourists, with an investment of $238 million into making the resort carbon neutral by 2020. If successful, then it will increase the resort's appeal to **ethical tourists**.

Terrorist attacks in 2015 have drastically reduced the number of tourists visiting the resort, which has eased the water supply issues but had negative effects on the socio-economic development of the region.

Now test yourself

TESTED

1. What is the difference between ecotourism and ethical tourism?
2. Using an example that you have studied, describe the steps taken to make the tourist industry more ethical.

Exam practice

'A debt-for-nature swap is the most effective environmental strategy to manage habitat and biodiversity.' To what extent do you agree? [8]